Tucholsky Wagner Zola Scott Sydow Freud Schlegel
Turgenev Fonatne Wallace
Twain Walther von der Vogelweide Fouqué Friedrich II. von Preußen
Weber Freiligrath Frey
Kant Ernst
Fechner Weiße Rose von Fallersleben Richthofen Frommel
Fichte Hölderlin
Engels Fielding Eichendorff Tacitus Dumas
Fehrs Faber Flaubert
Eliasberg Ebner Eschenbach
Maximilian I. von Habsburg Fock Zweig
Feuerbach Eliot
Ewald Vergil
Goethe Elisabeth von Österreich London
Mendelssohn Balzac Shakespeare Ganghofer
Lichtenberg Rathenau Dostojewski
Trackl Stevenson Doyle Gjellerup
Mommsen Hambruch
Thoma Tolstoi Lenz Droste-Hülshoff
Dach von Arnim Hanrieder
Verne Hägele Hauff Humboldt
Reuter Rousseau Hagen Hauptmann Gautier
Karrillon Garschin
Defoe Hebbel Baudelaire
Damaschke Descartes
Hegel Kussmaul Herder
Wolfram von Eschenbach Schopenhauer
Dickens Grimm Jerome Rilke George
Bronner Darwin Melville
Bebel Proust
Campe Horváth Aristoteles
Voltaire Federer
Bismarck Vigny Barlach Herodot
Gengenbach Heine
Storm Casanova Tersteegen Grillparzer Georgy
Chamberlain Lessing Langbein Gilm
Gryphius
Brentano Lafontaine
Strachwitz Claudius Schiller Kralik Iffland Sokrates
Bellamy Schilling
Katharina II. von Rußland Gerstäcker Raabe Gibbon Tschechow
Löns Hesse Hoffmann Gogol Wilde Vulpius
Luther Heym Hofmannsthal Gleim
Roth Klee Hölty Morgenstern Goedicke
Luxemburg Heyse Klopstock Kleist
Puschkin Homer Mörike Musil
Machiavelli La Roche Horaz
Navarra Aurel Musset Kierkegaard Kraft Kraus
Nestroy Marie de France Lamprecht Kind Kirchhoff Hugo Moltke
Laotse Ipsen Liebknecht
Nietzsche Nansen Ringelnatz
Marx Lassalle Gorki Klett Leibniz
von Ossietzky May
vom Stein Lawrence Irving
Petalozzi Knigge
Platon Pückler Michelangelo Kafka
Sachs Poe Kock
de Sade Praetorius Mistral Liebermann Korolenko
Zetkin

The publishing house tredition has created the series **TREDITION CLASSICS**. It contains classical literature works from over two thousand years. Most of these titles have been out of print and off the bookstore shelves for decades.

The book series is intended to preserve the cultural legacy and to promote the timeless works of classical literature. As a reader of a **TREDITION CLASSICS** book, the reader supports the mission to save many of the amazing works of world literature from oblivion.

The symbol of **TREDITION CLASSICS** is Johannes Gutenberg (1400 – 1468), the inventor of movable type printing.

With the series, tredition intends to make thousands of international literature classics available in printed format again – worldwide.

All books are available at book retailers worldwide in paperback and in hardcover. For more information please visit: www.tredition.com

tredition was established in 2006 by Sandra Latusseck and Soenke Schulz. Based in Hamburg, Germany, tredition offers publishing solutions to authors and publishing houses, combined with worldwide distribution of printed and digital book content. tredition is uniquely positioned to enable authors and publishing houses to create books on their own terms and without conventional manufacturing risks.

For more information please visit: www.tredition.com

The Teesdale Angler

R. Lakeland

Imprint

This book is part of the TREDITION CLASSICS series.

Author: R. Lakeland
Cover design: toepferschumann, Berlin (Germany)

Publisher: tredition GmbH, Hamburg (Germany)
ISBN: 978-3-8491-4868-3

www.tredition.com
www.tredition.de

Copyright:
The content of this book is sourced from the public domain.

The intention of the TREDITION CLASSICS series is to make world literature in the public domain available in printed format. Literary enthusiasts and organizations worldwide have scanned and digitally edited the original texts. tredition has subsequently formatted and redesigned the content into a modern reading layout. Therefore, we cannot guarantee the exact reproduction of the original format of a particular historic edition. Please also note that no modifications have been made to the spelling, therefore it may differ from the orthography used today.

PREFACE.

I find it requisite to say something by way of preface to the Tees-dale Angler, chiefly, because I wish it to be understood that my work, though bearing a local title, is intended as a help and guide to Trout fishers generally, especially those of Yorkshire, Durham, Westmoreland, and Cumberland.

To the extent of my ability, I have endeavoured to point out, and explain the various methods, means, and devices, natural and artificial, for taking Trout. The Artificial Fly List will I trust be found amply sufficient for most Anglers. I have only to add, that my treatise is the result of a considerable amount of practical Angling experience, extending over a period of upwards of 35 years, and the chief object I have in view will be accomplished, if the hints and instruction contained in it, tend to aid the diversion, and promote the amusement of those who wish to be proficient in the art of a pleasing and fascinating recreation.

R. LAKELAND.

CONTENTS.

THE TEESDALE ANGLER.

- *Salmo* — The Salmon.
- *Trutta* — The Trout.
- *Thymallus* — The Grayling.
- *Capito Seu Cephalus* — The Chub.
- *Salmonidæ* — Smelts.
- *Anguilla* — The Eel.
- *Various seu Phocinus* — The Minnow.
- *Cobitus Fluviatilis Barbatula* — The Loach.[1]

I deem a very brief notice of the above varieties of fish sufficient, — they have been described over and over again by much abler pens than mine, and I advise all those who are desirous of minute details, as to their conformation and habits, to have recourse to one of the published Histories of British Fishes,[2] indeed all the above fish and their varieties have been faithfully and naturally described in (I take it for granted) every angling book that has yet been published. As to Salmon, I need allude no further than observe (as every one knows that they are both ocean and river fish) that they afford, when plentiful, excellent sport to the angler, taking freely the Minnow, Worm and Fly, that they generally select the deepest pools of a river for their chief residence, but yet may be taken anywhere with the fly where there is three feet of water. They generally rise best about eleven o'clock in the forenoon, and three in the afternoon of a day. When there is a little wind stirring, if accompanied by rattling showers of hail or snow in the Spring, or heavy showers of rain in Summer, so much the more likely for sport.

Salmon fishing in every respect is similar in the *modus operandi*, to that of Trout, — requiring not more, if so much skill, but more nerve and patience with, of course, much stronger rod and tackle, and larger flies, and if you try worms, two large lob worms well scoured, should be put on the same hook, — you also require a Gaff for large fish. The best Salmon Flies for the Tees (which is by no means a good Angling river for Salmon) are the Dragon and King's Fisher, to be bought at most tackle shops, and a fly deemed a great killer made with a bright scarlet body, and wings from the black feather of a turkey.

THE TROUT.

The Trout almost every one knows, that the Trout is a delicious fish, beautiful and elegant in form and appearance. Trouts vary, being yellow, red, grey and white, the latter like Salmon, go into salt water. Trout spawn in the winter months, after which they become sickly and infested with a species of what may be denominated fresh water lice. In winter he keeps to the deep water; in spring and summer he delights in rapid streams, where, keeping his head up the water, he waits for his expected prey. There is no other fish that affords such good and universal sport, or that exercises the skill and ingenuity of the angler so much. The different modes by which to effect his capture are fully described under the different heads of fly trolling and bottom fishing. This fish (but seldom taken any great weight) abounds in the Tees and its tributary streams.

THE GRAYLING.

The Grayling is a beautifully formed fish, and affords the angler good sport—he is a much better-flavoured fish than the Chub, though not comparable to Trout. He delights in rapid streams, and during the Summer months is rarely found in deep water. The Grayling will take the same flies and bait as Trout—a little black fly is an especial favourite with him, but he will spring a long way out of water to catch a fly of any description which may be sporting above him. The Grayling spawns at the end of April and beginning of May.

CHUB, OR CHEVIN.

The Chub is a very timorous fish, utterly worthless as food except during the winter months. He frequents deep water, and loves shady places, where he can shelter under the roots of trees, &c. The Chub spawns in May and June. He is a leather-mouthed fish, so that once hooked you are sure of him; he struggles fiercely for a moment, then yields without further effort, and allows himself to be dragged unresistingly to land. He will take the same flies as the Trout, also all kinds of gentles, maggots and worms, especially small red worms; is fond of the humble Bee, Salmon Roe, and

Creeper; will take a variety of pastes, as old white bread moistened with a little linseed oil and made into small balls; old Cheshire cheese mixed with a little tumeric, and bullock or sheep's brains, also bullock's blood mixed with wheaten flour, and worked up to a proper consistency, are all good baits for Chub in the winter months. A Cockchafer with his wings cut off is also a very good bait for large Chub. When rivers are frozen, you may catch Chub by breaking a hole in the ice, the fish will come to the aperture for air, and, perceiving the bait, take it—your line need not extend to the depth of more than a yard. Observe that your paste balls are of consistency sufficient to adhere firmly to your hook, which should not be larger than a small May-fly hook, or two No. 3 fly hooks tied firmly together are much better.

SALMON SMELTS.

The growth of Salmon, as is well known, is so surprisingly quick, that Smelts from Ova deposited by Salmon during the Autumn and Winter months, will in some instances, by the first week in May, be found to weigh after the rate of five or six to the pound. They rise very freely at the fly, and afford the angler (who is fond of small fry), lots of sport, they are partial to streams, and also to a gaudy fly. Smelts will rise at almost any moderate sized fly, but the three most killing, are a small black fly, with scarlet or crimson silk body, black fly, ribbed with gold, or silver twist, golden plover's speckled feather from the back, and gold twist. They are also rather fond of a fly made from a partridge's breast feather, and body of crimson floss silk. The flies must be fastened upon small hooks not larger than No. 1. Few Smelts are to be seen after the second week in May. There is an old saying,

"That the first flood in May,

Takes all the Smelts away."

Salmon Trout, or Herling as they are called in Scotland, are a beautiful and elegantly formed fish, and rise very freely at common Trout Flies, these fish go into salt water.

THE PINK, OR BRANDLING.

The Pink is plentiful in the Tees and many of its tributaries, it is altogether a handsomer fish than the Trout, to which however in some respects it bears a strong resemblance. It is seldom taken above a quarter of a pound in weight. Is very vigorous and strong for its size, delights in rapid streams, takes the same baits and flies as the Trout, but when the water is low and the weather hot, is exceedingly fond of the maggot, or brandling worm. The Cad bait, with a little hackle round the top of the shank of the hook, kills well. The hackle should be Landrail, or a Mallard's feather dyed yellow, the latter for choice.

THE EEL

May be termed amphibious, for about the time oats run, he has been met with at considerable distances from water, and has even been detected in pea fields, gorged with the usual accessories to duck, to which in some respects he is so far analogous—that though a foul feeder he is excellent as an edible. He inhabits mud and sand banks, and also conceals himself under tree roots, stones and rocks. You may angle for him with Salmon Roe, a lob-worm or Minnow after a flood and before the water has subsided, but he is usually taken by night-lines, baited with lob-worms or Minnows. As I have before intimated, he is not nice, and will not refuse any kind of garbage. If you angle for him your tackle should be strong and leaded, so as to keep your line at bottom.

THE MINNOW.

The Minnow is in deep water during winter, and the shallowest of streams in summer; he is taken with a small red worm, or with young Cad bait. The Minnow bites freely in fine weather, and you may take almost as many as you please by angling for them. When the water is clear, they may be taken by means of a large transparent glass bottle, wide at the top of the neck but gradually narrowing, in fact a complete decoy; inside the bottle are red worms, and the bottle, to which is attached a string, thrown round the neck, is cast into the water; in a little time a shoal of Minnows surround the bottle, enter, and feast. When the bottle is tolerably full, a pull at the string brings bottle and Minnows to land.

THE LOACH

Is found underneath stones at the bottom of rivers and brooks, and also amongst gravel; it is a good bait for Trout and Eels. The Loach will bite freely at small red worms. The hook same as for Minnows.

THE BULL-HEAD

Though an ugly looking fish is good to eat; you may catch him with any small worms and small hook, he is found amongst stones and gravel.

ADVICE TO BEGINNERS.

Angling is such a popular recreation that professors of the gentle craft are to be found amongst all classes and conditions of the *Genus homo.* The disciples of glorious old Izaack—is not their name Legion? In early youth, fascinated with the capture of the tiny Minnow or glittering Gudgeon, the youthful Tyro is known in after years as the expert Salmon and Trout fisher. To become a really expert angler, requires a good deal of energy, perseverance, and activity, accompanied by a suitable amount of patience and ingenuity. In the fourth chapter of Waverly are the following observations, "that of all diversions which ingenuity ever devised for the relief of idleness, fishing is the worst qualified to amuse a man, who is at once indolent and impatient, such men's Rods are quickly discarded." My advice to those who are desirous of enjoying "the contemplative man's recreation," is that they undergo a probationary course, under the guidance of a competent professor. Three or four days of diligent observation employed in watching the manual operations of an instructor, would go far towards giving them a pretty good idea of how to set about catching a Trout with either fly or bait; indeed much more so than any written or oral instruction could convey. In fact if they are attentive spectators, they may soon acquire a fund of useful practical information, with which they may commence angling with a fair chance of success. Theory may be very good, but practice is much better, and will only make the complete angler. Good Rods, superb Flies, and the best of all kinds of tackle are of little use, if any, in the hands of a person who has not previously acquired some notion as to the proper application of them. Doubtless many a sanguine aspirant to piscatory fame, has, after an expensive outlay at a tackle shop, been grieviously disappointed when trying his luck in a celebrated Trout stream,—he discovers to his intense disgust and mortification, that the fish will "not come and be killed." Probably, and indeed most likely, he throws down his rod, votes fishing a bore,

"Chews the cud of bitter disappointment o'er,

Has fished his first and last, and so will fish no more."

The manual part of angling is one thing, the commanding success another, the latter cannot be effected to any extent without the sacrifice of time, perseverance and attention. It is however quite probable that a man may be quite happy and satisfied by the capture of a very small number of Trouts during a day's fishing, and I strongly advise all beginners to follow so excellent an example, waiting patiently "the good time coming." Observe, that fishing in a low water, where an angler has just preceded you, is the *ne plus ultra* of doing worse than nothing; by wading in a low water the fish are so scared that they take to their holds, and probably remain there for some hours.

VARIOUS USEFUL HINTS.

By keeping your tackle-book neat and tidy, you will always have your silks, hooks, lines, flies, &c. in their proper places. When the twine that holds your two-piece Rod together has been thoroughly wet, then when dry, and before using it again, wax well. If any portion of a Rod of three or more pieces is so fast at the joints that you cannot draw, then hold over the flame of a candle or by the fire, and then try, the result is generally satisfactory. Let your gut soften in the water before you commence fishing. Examine old stintings of gut and hair to see there are no flaws by wear and tear, if there are, repair, or discard altogether, carelessness in such matters always brings disappointment in the long run. See that the points of your hooks are sharp, and that the hooks are all right, as broken or crooked hooks are of course useless. Make it a rule to examine closely any place where you have had your book out dressing flies, &c., so that you leave nothing behind. If your flies or hooks are fast to any impediment which you cannot reach, don't pull like a savage, but go tenderly and cautiously to work; a release is often effected by a little time and patience; when the case is utterly hopeless, and a breakage becomes inevitable, then try to save as much of your tackle as possible. Never loose your temper because you loose your fish, let hope "whisper a flattering tale" for the next you hook. When you have hooked a fish, don't let him run if you can possibly help it, so as to slacken your line, if you do, you stand a chance of loosing him, as the sudden cessation of a strain upon the line frequently disengages the hold. If you want to discover what fish are feeding upon, open the first you catch, and then you will be able to judge correctly. Never strike a fish hard with the fly, either on gut or hair, if the latter, a breakage is almost sure to follow a violent jerk. Stormy, showery days in summer and sometimes in spring, are days on which you will generally take the best fish with the fly. After a flood, with a rising barometer, and not too much wind, expect good sport. If the fish do not like the worm after you have tried a few likely places, change for the fly, and if you do not succeed with that, wait twenty minutes or so, and probably you may then find them disposed to feed. Whenever you find fish shy in taking the worm, I mean when they will neither take it nor let it alone, pulling at it but not attempting to gorge it, strike either very quickly, as soon in fact as you perceive they have touched it, or what will generally answer much better, exchange for the fly. Sometimes, however, fish will take worm very well, although they may be seen rising freely at the fly. Cold dark days are not favourable for worm fishing, and in low water the worm is entirely useless on such days. Put your Minnows for trolling in tin cases, with partitions for each Minnow with a little bran in each, this method keeps them nice and fresh. Observe, that Loaches, if you can get them are tougher than Minnows, and quite as good if not better bait. Never buoy yourself up with the hope of having any diversion, either at top or bottom in an easterly wind. Also after a frosty night followed by a bright day, fly fishing need not be attempted with any chance of success. Put your worms when you are going to use them in a woollen bag in Spring, canvas in Summer. In May-fly season, if there comes a flood, go at the rising of the water and secure as many as possible, you will find them scarce afterwards. If, when fishing up water you meet an angler coming down, you had better wait twenty minutes before you try the stream recently fished. Guard against your shadow falling upon the water, at least as much as possible. If you purpose wading, be careful not to over-heat yourself during your walk to the water side. If, when the morning has been cloudy, and the fish have risen tolerably at the fly, should the sun appear about noon, coming out strong and likely to continue, you will find the fish cease to rise, and it is very probable that they will feed no more until evening. After a white hoar frost, either in the Spring, or further on in the season, fish rarely feed until the afternoon of that day, and not always then. When a thick mist rises from the water early on a Summer morning, fish will not feed until the vapour rising from the water has passed away. On stormy days try mostly that part of the water where there is the best shelter to be had.

ON FLY FISHING.

In *Thompson's Seasons* what an admirable description of Fly fishing! It is indeed inimitable: it charms an angler by its vivid and truthful deliniation, and after reading it, makes him long "to increase his tackle, and his rod retie." Of all the devices for taking Trout, fly fishing is decidedly the most pleasant, ingenious and amusing, and where fish rise freely, there is nothing comparable to the artificial fly, as a means to an end, in the shape of filling a pannier. The quick eyed Trout, is completely deceived by a cunning fabrication, the inanimated thing of feathers, silk and fur, so closely resembles the natural fly, that he rises and seizes upon it for a real living fly — But ah! too late, the little monster (for he is one in his way) feels the treacherous hook, "indignant at the guile," he springs aloft, makes for his well known hold, or resting place, exhausts his strength in the unequal contest, and floats almost lifeless into the landing net held out for his reception. He has fallen a legitimate prize to the skill of his captor, who has only to extract the hook from his gills, before he again makes another light and deadly cast. Thus fish after fish is deposited in his nicely woven pannier, and on he goes rejoicing, carefully trying his favourite streams, until the weight upon his shoulder, unmistakably intimates, that it is time to be homeward bound. In fly fishing, the best plan is to cast your line athwart the stream, by pulling it against it; your flies probably show to more advantage, yet you will not take so many fish, as by throwing up or across the stream, the reason is obvious, the current somewhat retards the progress of the fish in the act of rising, and thus it happens that they so frequently come short of the hook. There is also another consideration, your fly coming down or athwart the water is more natural, and fish observe it sooner coming down, than a fly pulled up stream, because fish when on the feed, invariably lay with their heads up water.

> "With pliant rod athwart the pebbled brook,
>
> Let me with judgment cast the feather'd hook,
>
> Silent along the grassy margin stray,
>
> And with a fur wrought fly delude the prey."
>
> Gay.

In log, or still water fishing, make as fine and light casts as you possibly can. If you see a fish rise, throw your flies about a foot above him, and then let them gently float over the place where he rose. In stream fishing, have a quick eye, and ready hand, and strike immediately you perceive the fish to have risen at your fly; and observe that if you have the luck to hook two Trouts at the same time, net the one lowest down your line first, for should a novice inadvertently attempt to net the one upon the higher fly, he will very probably loose them both. The heads and tails of streams are favourite resorts of Trout, and ought to be carefully and diligently fished; but as a general rule, wherever you see a fish rise, have a try for him. In the Spring and Autumn, your diversion with the artificial fly is much more certain than during the Summer months, but even then there are certain days, (especially if the wind be Easterly), that they will not take even the natural fly, and I have on such days seen thousands of flies on the water, yet scarcely a fish on the move. When the fish rise freely at the natural fly, and also rise, but do not take those you offer, you may safely conclude your fly is not what suits, so try them with something different. The best plan is to catch the natural, and make the artificial fly as close a copy as possible, for the nearer you approach to nature the greater in all cases is your chance of success. And here, in concluding this chapter on Fly Fishing, let me advise every angler to make or learn to make his own flies; by so doing he will never be at a loss for a fly to suit the fickle Trout. Really, many of the flies from the tackle shops look neat and gaudy enough, but like Hodge's razors, are they not made to sell? When a man makes a fly for himself, he makes, I take it, to kill.

THE ANGLING MONTHS.

March.—During this month the fells and hills of north Yorkshire and Durham are frequently capped with snow, which, dissolved by the increasing power of the sun, fills rivers and brooks with what is usually termed snow broth, which, accompanied with chilling east or north-east winds, effectually retard angling operations. Trout however keep gradually improving in condition, and from the middle to the end of the month will, under the influence of a kindly atmosphere, rise tolerably well at the fly during the middle of the day. The worm is also taken in brooks after rain. But as a fly fishing month, March seldom affords, in the north of England at least, any good or certain diversion. In the face however of all obstacles, some really keen hands will wet their lines, and if the weather is at all genial, may succeed in taking a few fish.

The advent of our annual visitor, the swallow, indicates, or nearly so, when fly fishing commences with some certainty of sport;[3] you will observe but few flies on the water, (and consequently no inducement to fish to be on the look out), before those great insect killers appear. The principal flies for the month, are the March Brown, the Blue Dun,[4] and small Black, or Light coloured flies.

Some anglers fish with four flies upon a stretcher, I much prefer three, and never, except for Lake fishing, use more—a stretcher for three flies should consist of about a yard and a half of either gut or hair. What are termed water knots are the best for tying your gut or hair together, the tighter they are drawn the faster they become. Every angler is no doubt partial to some particular flies, and probably he will have no great difficulty in selecting his favourites from the copious lists given in the Teesdale Angler; but for the benefit of those anglers who have not had much experience, I beg to observe that they should never have three flies at once on their stretcher, that closely resemble each other. In the Spring the Blue, Brown and Dun Drakes are certain killers, and as for hackle flies, if they select the Brown, Blue and Black, they will do well. During the Summer months there is such a great variety of feed upon the water that it is difficult, nay, almost impossible to give any certain rule, because the set of flies that kill well one day, may be rejected the next. I may however venture to affirm, that one dark and two light flies are the most likely, either as regards hackle or winged flies. By catching the natural fly, you will never be at a loss either in Spring or Summer, as to the colour of the silk you require for the body of your fly. In Summer when the midges are on, use the Black, Blue and Dun midges, and when they disappear, try the larger flies.

April. — The month of sunshine and showers is generally, and especially towards the latter end of it, most favourable for angling; in fact if the water is in order, and the weather temperate for the season, it is the very best fly month in the year. Trout are now sure to rise well and freely at the fly. Every day between the hours of eleven and three o'clock the feed is on the water. The fish, full of life and motion, are hungry and voracious, and in full pursuit of the Dun or Brown Drake, which any gleam of sunshine brings on the surface of the water. The Blue Dun (a better fly than the Brown for cold stormy days), and the Grannam, or Green Tail, are frequently on at the same time, and it is a pleasant sight to anglers to see thousands of these flies settling on the water, and the fish rising at them in all directions. During these feeds I venture to predict that any person who has suitable flies, and who can manage to make a tolerable light cast, cannot well miss taking some fish. With respect to the Grannams, you may on bright mornings begin to fish with them as early as six o'clock, and again after the large Browns have disappeared, I mean for that day. If you commence fishing, say any time between six and eleven a.m., use the small flies, viz., the Grannam or Green Tail, the small Blue Dun, and Black Flies, dressed on No. 2. hooks. — During this month (April) it is frequently so cold that to dress a fly by the water side is almost an impossibility, or at least a matter of some difficulty, therefore, always be provided with a supply, ready for use when wanted. I also strongly recommend fine round Gut in preference to Hair at this season, on account of the size and weight of the large hooks on which the Brown Drake requires to be dressed; and which Hair will not retain so safely as Gut; and also, though you may probably rise more fish with Hair, yet taking the breakages you are liable to by using it, and the loss and hinderance you suffer thereby, especially if broken in the midst of a feed, which perhaps does not last above a quarter of an hour, taking these matters into consideration, I have long since arrived at the conclusion that Gut is much better for Spring fishing than Hair. But in the long Summer day, when your fingers are not benumbed with the cold, and you can dress flies or repair and arrange your tackle at your pleasure or convenience, then, when the water is low and fine, there is nothing comparable to strong, fine round Hair, it falls much lighter than Gut on the water, and therefore, for log or still water is much superior. But really good Hair for angling purposes is exceedingly difficult to meet with, and if you use inferior, many losses and disappointments are sure to occur. Good Hair has the advantage over Gut in these respects,—it is sooner wet, falls lighter on the water, and is free from that glistening and shiny quality which detracts so much from Gut, and which no staining will entirely obliterate; it wears out by use in a great measure, but having come to that point, cannot be depended upon, and if you lay it aside for any length of time when in that state, you will find, if you attempt to make use of it, that it is utterly worthless. The shaved Gut is good, but expensive. The best I ever purchased was at Rowell's, at Carlisle.

May, "charming, charming May," is generally a delightful Angling month, for if the water is in order, good diversion may be had almost every day. A great variety of flies now make their appearance at which the Trout rise very greedily, full of life, vigour and activity, they roam everywhere after their prey, and scarcely a fly settles upon the water but falls a victim to the quick eyed and hungry fish. Trolling, and worm fishing become now very good, and it is advisable to fish with either one or the other in the early part of the day. When the flies have not made their appearance, and before fish rise of themselves, it is of little use trying the fly, it is only labour lost, *"to call spirits from the vasty deep, who will not come when you do call for them."* Indeed, on the best of fishing days, there are some half hours when a man who understands what he is about, will lay down his rod, because he knows the fish have done feeding for a time, and that flogging the water to no purpose may be exercise, but not sport. In this leisure half hour then, let the angler smoke, eat, examine his Tackle, or lay out and admire his fish, this last way of killing time, brings to my recollection the lines of Wordsworth,—

"He holds a small blue stone,

On whose capacious surface is outspread,

Large store of gleaming crimson spotted trouts,

Ranged side by side in regular ascent,

One after one still lessening by degrees,

Up to the Dwarf that tops the pinnacle,

The silent creatures made a splendid sight together thus exposed;

Dead, but not sullied or deformed by death,

That seemed to pity what he could not spare."

Wordsworth.

June, loveliest of the Summer months, introduces to the notice of anglers a large and daily increasing number of the insect tribe; "variety may be charming," but the most expert and knowing of anglers will now occasionally be somewhat puzzled in making a selection of flies adapted to suit the capricious whims, or fastidious appetite of the Trout, now in their prime, fat, strong, and somewhat satiated by a succession of dainty morsels. Now is the time to rise with, or rather indeed before the lark, and try your luck with the creeper and stone fly, you may begin to fish with either as soon as you can see to put them on the hook, and always bear in mind that the early morn is the best part of the day for these baits, you also have a good chance again in the evening, but in the middle of the day they are, upon the whole, but indifferently good; and the small fly will generally be found to answer better,—and frequently the worm proves destructive when the day is hot, and the water low. It is a good plan to procure your May-flies and creepers during the day or evening preceding that on which you intend using them, searching for them in the morning when you want to fish is not quite pleasant. You may do a great deal of execution with the small flies just now. Trout glutted with the May-fly and creeper, take them well on cloudy and windy days. Should rain fall at this season, after the water has been low for some time, Trout will take a minnow exceedingly well.

July.—The scorching suns of Summer are upon us, and the vivid rays of the great luminary have a powerful effect upon all creatures, and upon the finny tribe in particular. The water during this month is often so low and fine, that artificial fly fishing is labour in vain, and provided it is not, fish have become so shy and cautious in the selection of their food, that it is a difficult matter to offer for their acceptance anything artificial which they will take freely. A well scoured worm, maggot, gentle or natural fly, offered to them in an artistic way, seldom however fails to attract their notice, — of natural flies the Flesh Fly is the best. Evening fishing, towards dusk, with the brown and white Moths, and also with the white Bustard, may be pursued with success; you may fish with the Bustard (which you will find performing aerial evolutions over the meadows in a fine evening) the whole night through, and though perhaps you cannot see the fish (which is generally a good one) rise, you must always strike quickly, yet gently, when you feel him — use a May fly hook. If you can find any May-flies, the fish will now take them again very greedily, during the last fortnight of this month very few fish can be taken under any circumstances with either natural or artificial flies, the fish are too fat and indolent to take the trouble to rise. A well scoured maggot on a bright hot day tempts them best, they will take that when flies and all other baits have proved a failure.

The Spring and Autumn fishing are easy enough, but the Summer tests the Angler, — and

> "Who then his finest skill and art must ply,
>
> And all devices, natural and artificial try,
>
> For now the Trout becomes an epicure indeed,
>
> And only on the daintiest baits and flies will feed."

August. — The same Flies as in July, with the addition of the little Red and Black Ant Flies, which usually appear about the 10th or 12th of this month; observe that from the 12th to the end of the month, fish take the fly much better than they have done — they are on the move again.

September and October. — Use the same Flies as in Spring, the willow fly in September must however be added to the list of Blues, Duns, and Browns. About the middle of October I deem it high time to lay aside the Trout Rod, let "the gentle angler" for a brief space bid adieu to his favourite piscatorial haunts, in doing so perhaps he may call to mind the farewell of the Tyne fisher to his favourite streams, from a work printed for Emmerson Charnly, at Newcastle, in 1824.

> Mine own sweet stream! thy rugged shores are stripped of all their vesture sheen,
>
> And dark December's fury wars where grace and loveliness have been,
>
> Stream of my heart! I cannot tread thy shores so bleak and barren now,
>
> They seem as if thy joys were dead, and cloud with care my anxious brow.

In reference to the above, I must observe that very few anglers will think of fishing during the winter months; at the conclusion of the second week in October, the Trout Rod ought to be carefully stowed away. The angler should by all means refrain from killing Trout so close upon the spawning season, besides they are becoming as food quite worthless. Truly "Othello's occupations gone."

NATURAL FLY FISHING.

THE STONE FLY.

The Stone Fly is invariably converted into the May-fly, by anglers who fish the Tees and its tributary streams; but the actual and properly named May-flies are the Green and Yellow Drakes, which do not appear upon our Teesdale waters. If the weather has been warm, and the water low, May-flies (for by so calling them I shall be best understood), may be found the last week in May, or at all events in the beginning of June, some indeed, but very few may be seen as early as April, and as late as September. This fly is easily found, his whereabouts indicated by his old coat, or husk, which he has discarded, and left on the outside of his mansion, which is generally a flat stone near the edge of the water. This fly is generally but an indifferent killer in the middle of the day, mornings and evenings, (when not glutted and the weather propitious), Trout take it with avidity, provided there has been no frost during the night, and the water is free from the steaming sort of mist prevalent about this season. You may begin to fish with the May-fly as soon as you can see to put the fly on the hook, the earlier you commence the better chance of large fish, especially if the water is clear, and very low, or even moderately so. In fishing with this fly, have your cast line light and strong, tapering gradually to the end, to which attach about three-quarters of a yard of fine round Gut, the best you can procure, on which tie your hook which must be at least a size larger than the Palmer hook; arm this hook with a strong pig's bristle, which must lay on the back of the hook, protruding a short way over the top of the shank. In putting on the fly, insert the point of the hook under the head of the fly, passing through the body, bring it out underneath the tail, then take and press the fly upwards over the head of the bristle on your hook, bringing it so far down that it may pass through the back, behind the head of the fly, then set to work by throwing your fly into rapid streams, eddies caused by rocks, or other impediments; cast your fly always up and let it come down the stream floating on the surface of the water in a natural and easy way; if a fish rises and does not swallow it, do not pull your fly away, the odds are he will follow and take it, his motive I suppose in the first instance being to disable; however when Trout are fairly glutted with the May-fly, they may rise, but will not even touch it. When a fish has seized your fly, do not strike too hard or hastily, numbers of fish are lost by doing so, let them always turn their heads either in stream or log water before you strike. On dark cold windy days, during the May-fly season you will find the small fly a much better killer than the May-fly. On bright and very hot days a well scoured Brandling Worm or Creeper may be used to advantage, after your morning's fishing with the May-fly is done, for on such days the artificial fly is entirely out of the question. A Bullock's horn with a few small holes bored in it, is perhaps the best and handiest thing you can put your flies into.

Observe that the Alder or Orl fly, is a capital killer when the May-fly is on. Who shall say that the May-fly short as is its life, has not undergone all the vicissitudes of a long and eventful life, that it has not felt all the freshness of youth, all the vigour of maturity, all the weakness of old age, and all the pangs of death itself?

TO THE MAY-FLY.

Thou art a frail and curious thing,

Engender'd by the sun,

A moment only on the wing,

And thy career is done.

Thou sportest in the evening beam,

An hour, an age to thee,

In gaity above the stream,

Which soon thy grave shall be.

Borton Wilford.

THE FLESH FLY.

The Flesh-fly, when the water is low and clear, is one of the most alluring flies that can be offered to the Trout, but great skill, care, and judgment are requisite in the use of it; in the hands of an expert angler, on a close hot day during the month of July, it is a sure and certain adjunct towards filling a pannier. The fish will take it when they will not look at an artificial; you will take as large fish with it as are to be had with any kind of fly, either natural or artificial. The flies are easily procured in shady places, in woods or fields, where cattle and horses have left recently made soil. After having struck them with a bundle of twigs and killed, or stunned, as many as will answer your purpose, put them into a horn, or anything suitable, so that they do not escape. Your cast line must be of a length proportioned to the size of the river or brook where you fish, as a general rule (if you wade in the water), about a little longer than the length of your rod, — let your cast line be exceedingly fine, and have attached to it three-quarters of a yard of the finest round silk-worm gut, — your hook should be No. 2, put your fly on by inserting the point of the hook under the head of the fly, and running it through the body, bringing it out at the tail — you need not make above two or three casts at a place, and follow the same rule as with the May-fly, viz., to let the fish turn his head downwards before you strike. Streams are the likeliest places where they have not time to scan the fly, in that curiously suspicious and shy manner in which they generally come to it in smooth water. However when they are in the humour they will take it anywhere if you can only contrive to keep out of sight, *hie labor hoc opus est*; this is the trouble and difficulty in a low water; and note, it is not worth while attempting to fish with the Flesh Fly on cold windy days, let the water be in ever such fine condition. Trout take this fly best when the temperature ranges somewhere about seventy Farenheit. This fly is often taken when the May-fly is refused.

THE COW DUNG FLY.

The Cow Dung Fly is a good and enticing fly, it is easily procurable, as its name intimates, on foil left by cattle: if the water is low and clear, with a brisk wind stirring, you may use it advantageously, because the wind usually carries great quantities of them upon the water, which induce the fish to rise. These flies are found from May to October; fish with them in the same way as the Flesh Fly; a No. 2 hook is quite large enough for them. Wherever you see a fish rise, when fishing with this or the Flesh Fly, you may count upon him as your own four times out of six, if you only contrive to make a light and dexterous cast, over the place where you observe the fish rise. Dapping or Dibbing, or perhaps more properly Dipping, — this is another method of using the natural flies, and a very killing way too; your rod for this fishing must be of a good length, with a stiff top; your line composed solely of good, fine, strong gut, must be about but not less than a yard in length, — put your flies on the same sized hooks, and after the same way as you are directed to adopt in the other method where a longer line is used. Having stationed yourself out of sight, behind a bush, tree or rock, let your fly drop gently on the surface of the water, keep lifting and letting it fall so as just to cause the slightest perceptible dimple on the water, and if there is a fish at all hungry in your locality, you are pretty sure to have him. If a good fish is hooked, let your winch line go, because he will struggle furiously when he feels the hook, and the hold might give way, provided you were too hasty and anxious to land him. In dibbing, almost any kind of fly will answer. The day suitable for this should be warm, and the water rather low and clear.

LIST OF PALMER FLIES FROM MARCH TO OCTOBER.

The following list of flies will take fish in all Trouting streams of Yorkshire, Durham, Northumberland, Cumberland and West-moreland.

MARCH.

Dark Brown.
Great Whirling Dun.
Early Bright Brown.
Blue Dun.
Little Black Gnat.

APRIL.

Dark Brown.
Violet Fly.
Little Whirling Dun.
Small Bright Brown.

MAY.

Dun Cut.
Stone Fly.
Camlet Fly.
Cow Dung Fly.

JUNE.

Stone Fly.
Ant Fly. Little Black Gnat.
Brown Palmer.
Small Red Spinner.

JULY.

Orange Fly.
Wasp Fly
Black Palmer.
July Dun.

AUGUST.

Late Ant Fly.
Fern Fly.
White Palmer.
Pale Blue.
Harry Long Legs.

SEPTEMBER.

Peacock Harl.
Camel Brown.
Late Badger.
September Dun.

OCTOBER.

Same Flies as in March.

It is best to make your Flies in a warm room, or in warm weather out of doors,—your silk will then wax kindly, which is of great consequence in making Flies.

The three best winged Flies for Spring, are the Red Fly, Blue, Dun and Brown.

The three principal Flies for Autumn are the little Whirling Blue, Pale Blue, and Willow Fly.

February.—Red Fly.

March.—Red Fly, Dun Fly and Brown Drake.

April.—The same as March with the addition of the Grannam or Green Tail, and the Spider Fly.

May.—The Black Caterpillar, the Little Iron Blue, the Yellow Sally Fly, the Oak Fly and the Orl Fly.

June.—Sky Coloured Blue, the Cadiss Fly, the Blue Gnat, Large Red Ant Fly, Black Ant Fly, Little Whirling Blue, Pale Blue.

July.—Some of the same Flies as June, with the addition of the Wasp Fly, Black Palmer, July Dun, and Orange Fly.

August.—Small Red and Black Ant Flies, Willow Flies.

September.—Pale Blues, and Whirling Blue.

October.—Same as March, with the addition of the Dark and Pale Blues.

March.—1. The Dark Brown—dubbed with the brown hair of a cow, and the grey feather of a Mallard for wings. 2. The Great Whirling Dun—dubbed with squirrels fur, for wings, grey feather of mallard. 3. Early Bright Brown—dubbed with brown hair from behind the ears of a spaniel dog, wings from a mallard. 4. The Blue Dun—dubbed with down from a black greyhound's neck, mixed with violet coloured blue worsted, wings pale part of a starling's wing. 5. The Black Gnat—dubbed with black mohair, the wings of the lightest part of a starling.

April.—1. The Dark Brown,—brown spaniel's hair mixed with a little violet camlet, warp with yellow silk, wings, grey feather from mallard. 2. The Violet Fly—dubbed with dark violet stuff, and a little dun bear's hair mixed with it, wings from a mallard. 3. The Little Whirling Dun—dubbed with fox cubs fur, ash coloured, ribbed about with yellow silk, wings a pale grey feather from a mallard. 4. Small Bright Brown—dubbed with camel's hair, and marten's yellow fur mixed, wings pale feather of a starling.

May.—The Dun Cut—dubbed with brown hair, a little blue and yellow mixed with it, wings, woodcock, and two horns at the head from a squirrel's tail. 2. The Stone Fly—dubbed with dun bear's hair, mixed with a little brown and yellow camlet, so placed that the fly may be yellower on the belly and towards the tail than any where else, place two hairs from a black cat's beard, in such a way that they may stand upright, rib the body with yellow silk, and make the wings very large from the dark grey feathers of a mallard. 3. The Camlet Fly—dubbed with dark brown shining camlet, ribbed over with green silk, wings, grey feather of a mallard. 4. Cow Dung Fly—dubbed with light brown and yellow camlet mixed, or dirty lemon coloured mohair with the hackle of a landrail.[5]

June.—1. The Ant Fly—dubbed with brown and red camlet mixed, wings, starling's feather, pale. 2. Little Black Gnat—dubbed with black strands from an ostrich, wings, light feather from underneath starling's wing. 3. Brown Palmer—dubbed with light brown seal's hair, warped with ash coloured silk and a red hackle over the whole. 4. The Small Red Spinner—dubbed with yellow hair from behind the ear of a spaniel, ribbed with gold twist, a red hackle over the whole, the wings from a starling.[6]

July.—1. Orange Fly—dubbed with brown fur of a badger, warped with red silk, wings from dark grey feather of mallard, with a head made of red silk. 2. The Wasp Fly—dubbed with brown bear or cow's hair, ribbed with yellow silk, and the wings of the inside of starling's wing. 3. The Black Palmer—dubbed with black copper coloured peacock's harl, and a black cock's hackle over that, wings, blackbird. 4. The July Dun—dubbed with the down of a water-mouse, mixed with bluish seal's fur, or with the fur of a mole, mixed with a little marten's fur, warped with ash coloured silk, wood-pigeon's wing feather for wings.—A good killer.

August.—The Late Ant Fly—dubbed with the blackish brown hair of a cow, warp some red silk in for the tag of the tail, the wings from a woodcock. 2. The Fern Fly—dubbed with the fur from a hare's neck, which is of a fern colour, wings dark grey feather of mallard. 3. The White Palmer—dubbed with white peacock's harl, and a black hackle over it. 4. The Pale Blue—dubbed with very light blue fur, mixed with a little yellow marten's fur, and a blue hackle over the whole, the wings from a blue pigeon.—A very killing fly. 5. The Harry Longlegs—dubbed with darkish brown hair, and a brown hackle over it, head rather large.

September.—The Peacock Harl—dubbed with ruddy peacock's harl, warped with green silk, and a red cock's hackle over that. 2. The Camel Brown—dubbed with old brownish hair, with red silk, wings dark grey feather from mallard. 3. The Late Badger—dubbed with black fur of a badger or spaniel, mixed with the soft yellow down of a sandy coloured pig, wings dark mallard. 4. The September Dun—dubbed with the down of a mouse, warped with ash coloured silk, wings feather of a starling.

October.—Same as March.

As I never fished for Trout in November, I attempt no list of Flies for that month. From Michaelmas to the middle of February, all anglers should refrain from killing Trout.

Moths Brown and White for Evening Fishing.—The Brown—from the feathers of a brown Owl, dubbed with light mohair, dark grey Cock's hackle for legs, and red head. White Moth—strands from an Ostrich, wings from a white Pigeon, a white hackle for legs, and a black head.—Hooks No. 2 or 3. Good killers at dusk on a Summer's evening.

LIST OF HACKLE FLIES FROM FEBRUARY TO NOVEMBER.

February.—Small black flies, made from Starling's breast or Black bird, with black or purple silk—hook No. 1. Inside and out of Woodcock's wing and yellow silk. Plover's breast or Dottrel's wing feather and yellow silk—hooks No. 1 or 2; red Cock's hackle and yellow silk.

March.—Inside of Woodcock's wing and yellow silk, No. 2 hook. Dark Woodcock, and dark orange silk, No. 2 hook. Dottrel and yellow silk, No. 2 hook. Dark Snipe and crimson silk, No. 2 hook. Dark Snipe and purple silk, No. 1 hook.

April.—Woodcock's as for March. Inside of Woodcock's wing and yellow silk, No. 2 hook. Freckled Snipe and yellow silk. No. 2 hook. Dark Snipe and crimson silk, No. 2 hook. Dottrel and yellow silk,—inside of Snipe's wing, and pale yellow silk,—hooks No. 2.

May.—All the above April flies are taken, also, Partridge's breast and yellow or crimson silk, very light Dottrel's or plover's breast and fawn coloured silk, Blackbird and purple silk, Blackbird and dark crimson silk, sea Swallow and primrose silk, inside of Woodcock's wing and crimson silk—hooks, 1 or 2 according to water.[7]

June.—Most of the above, to which add Dottrel and orange silk, Plover and light orange silk, dark Snipe and orange silk, Freckled Snipe and orange silk, freckled Snipe and crimson silk. Hooks No. 1 or two according to size of water. Dottrel's breast and yellow silk,—Hooks No. 1.

July.—Many of the above, with Sandpiper and yellow or purple silk, Plover's breast and crimson silk Wren's tail and orange silk, Dottrel and bright scarlet silk; Plover's back feather with gold twist and orange silk, Landrail and bright red silk, dark Snipe and sky coloured blue silk.—Hooks No. 1 or 2 at discretion. If the water is very clear, use hooks as small as possible.

August.—Some of the July flies for the first fortnight, with dark Snipe and green, Snipe's breast and purple silk, Dottrel and black silk, Landrail and red silk, dark Snipe or Starling's breast and red silk, Grouse hackle and bright scarlet silk.—Hooks 1 and 2 according to water.

September.—Some of the August Flies, with Landrail and yellow silk, pale blue from sea Swallow and primrose silk, pale blue from ditto and crimson silk,—Hooks 1 and 2.

October.—Inside of Snipe's wing feather and yellow silk, Woodpigeon's feather and pale yellow silk, dark outside feather of Snipe's wing and crimson or orange silk, outside feather of Dottrel's wing and yellow silk—hooks No. 1 or 2.

November.—Same Flies as February.

The Blue, Black and Dun Gnats are at times on the water from May to August, and when the fish are taking them they generally refuse the larger flies.

The Blue Gnat may be made thus: A blue feather from a Titmouse's tail for wings, body from pale blue floss silk, on a cypher hook, which means the smallest hook made; or the wings may be had from Heron's plumes, with same or primrose silk.

Black Gnat—Starling's breast and black silk, cypher hook; or black Ostrich strand and inside wing feather of Starling for wings.

Dun Gnat—from inside wing feather of a Landrail and fawn coloured silk—cypher hook.

Observe, that you may put more feather on your hackle flies in the Spring than in the Summer; when the water is low and clear, a very small quantity of hackle is sufficient, and it should by no means descend much, if any, below the bend of the hook.

In low waters, except when the blue, dun and brown drakes are on, the hackle flies will generally be found to kill better than the winged flies.

REMARKS ON THE MARCH BROWN OR DUN DRAKE.

The March Brown is well known to all anglers as a fly to which they are chiefly indebted for the greatest portion of their sport in the Spring, commencing as its name indicates in March, and continuing the whole of April and into May. They appear on the water each succeeding day about eleven in the forenoon, and retire about half-past two p.m. Few rivers or brooks produce March Browns that are exactly alike;—I mean with regard to the same shade of colour, even in the same river there are frequently darker and lighter flies. For the lighter one I recommend the hen pheasant's or brown owl's wing feather, dubbed with hare's ear and yellow silk; for the dark, the tail feather of a partridge, a brown red hackle underneath the wings, and dark orange silk, or a woodcock's feather for wings, and a dark red hackle with dark orange silk,—kills exceedingly well. When the water is low and fine, I consider your chance of killing fish far greater with two, than three of the large spring flies. If you put the brown, and blue dun on your stretcher, three quarters of a yard apart, you will find your cast will be much lighter with the two than three; this plan also holds good in reference to hackle flies, provided that you know what the fish are taking.

SELECT LIST OF VERY KILLING FLIES, BOTH PALMERS AND HACKLES.

If these flies do not answer, it is very rare that you will succeed with any other. They are suitable for all the rivers and brooks of Yorkshire, Durham, Westmoreland and Cumberland; about thirty years experience has convinced me of their entire excellence, and probably the ingenuity of man cannot devise any to supersede them.

Palmers for March, April, and first week in May,—The March Brown or Dun Drake,—The Blue Dun,—Early Bright Brown.

May.—The Dun Cut,—The Cow Dung Fly, and also the March Brown and Blue Dun are on the waters in late seasons to the middle of the month.

June.—Little Black Gnat,—The Brown Palmer,—Little Red Spinner—and Alder Fly.

July.—The Wasp Fly,—Black Palmer,—July Dun.

August.—The Late Ant Fly,—The Pale Blue.

September.—The September Dun,—The Camel Brown and Willow Fly.

October.—Blue Dun, Pale Blue, and Dun Drake.

Note.—If there are no Flies on the water when you begin to angle, try a Palmer till you find what Flies the fish are taking. One Palmer and two small hackle Flies on your stretcher give a tolerable good chance.

LIST OF HACKLES AND SILKS TO SUIT.

(GOOD KILLERS.)

For March and April.—Dark Snipe and crimson silk,—Dark Snipe and Purple silk.—Hooks No. 1 and 2.—Outside feather of Woodcock's wing and dark orange silk.—Inside feather of Woodcock's wing and yellow silk.—Dottrel's back or neck feather and yellow silk.—Hooks No. 2 or 3.

May.—Inside and outside feathers of Woodcock's wing, with orange and yellow silk,—Starling or Blackbird's breast and black silk,—Freckled Snipe and yellow silk,—Dark Snipe and crimson silk.—Hooks No. 1 and 2.

June.—Blackbird and orange silk,—Plover and orange silk.—Dottrel's breast and yellow silk,—Freckled Snipe and crimson silk,—Partridge's breast and crimson or yellow silk,—Dark Snipe and yellow silk,—Freckled Snipe and orange silk,—Sandpiper and purple or yellow silk.—Hooks No. 1 or 2.

July.—Light Dottrel and scarlet silk,—Inside of Landrail's wing and yellow silk,—Blackbird and dark red silk,—Feather from neck of a Grouse and scarlet silk,—Plover's breast and bright yellow silk,—Sandpiper and purple silk.—Hooks No. 1 or 2.

August.—Most of the July hackles for the first fortnight, to which add dark Snipe and green silk,—Snipe's breast feather and purple silk,—Dottrel and black silk,—Landrail and red silk.—Hooks No. 1 and 2.

September.—Some of the August Flies with Landrail and yellow silk,—pale blue from Sea Swallow's wing and yellow or primrose coloured silk,—pale blue from Sea Swallow and crimson silk.

October.—Same as March,—with inside of Snipe's wing and yellow silk,—Woodpigeon's feather and yellow silk,—Dottrel and pale yellow silk.—Hooks No. 1 or 2.

I deem *November* like February, not worth a list.

A LIST OF FLIES THAT WILL, OR AT LEAST ARE LIKELY TO KILL, IN ALL TROUT STREAMS.

FLIES FOR MARCH.

1. Dark Blue,—one of the earliest. 2. Olive Blue,—March and April,—a good Fly in cold weather. 3. Red Clock,—April and March. 4. Little Brown,—March and April, the dark first, then the lighter,—good on warm days. 5. Blue Midge,—early in Spring and late in Autumn. 6. Great Brown, or March Brown,—March, April and first week in May. 7. Yellow Legged Blue,—from the latter end of March to the end of April, on cold days, particularly in April.

FLIES FOR APRIL.

1. Dark Blue,—yellow or Dun Midge from middle of April to middle of May. 2. Spider Legs,—end of April and May,—kills best in a wind. 3. Land Fly,—end of April till towards the end of May 4. Green Tail or Grannam, from six in the morning till eleven again in the evening, when the Browns are off. 5. Ash Fly,—from April to the end of June,—a good killer on windy days.

FLIES FOR MAY.

1. Grey Midge,—the latter end of April and all Summer. 2. Yellow Sally Fly,—all May. 3. May Brown,—latter end of May till latter end of June. 4. Pale Blue,—from middle of May and all through June,—good in the evenings. 5. Yellow Fly,—the greatest parts of May and June,—kills best on cold windy days. 6. Little Stone Blue,—from the middle of May till the Autumn. 7. May or Stone Fly,—if the weather is genial, the last week in May, and continues through June.

FLIES FOR JUNE.

1. Hawthorn Fly,—all June. 2. Little Dark and Pale Blue,—the dark during the middle of the day, the light in the evening. 3. June Dun,—about the middle of June,—suits showery weather. 4. Twitch Bell,—continues till the middle of July,—best in the evening,—Stone fly.

FLIES FOR JULY.

1. Little Olive Blue,—the greater part of July and August 2. Black and Red Ant Flies,—in July, August and September. 3. Little Blue,—July and August,—best in the middle of the day.

FLIES FOR AUGUST.

1. August Brown,—comes on about the latter end of July, continuing through August and till the middle of September. 2. Light Blue,—August, September and October,—a capital Fly on cold days. 3. Orange Stinger,—hot days in August. 4. Grey Grannam,—showery days in August and September.

FLIES FOR SEPTEMBER.

1. Light Olive Blue. 2. Small Willow Fly. 3. Large Willow Fly,—September and October.

FLIES FOR OCTOBER.

1. Blue Bottle and House Fly. 2. Small Olive Blue. 3. Dark Grey Midge.

HOW TO DRESS THE ABOVE.

FOR MARCH.

1. Dark Blue,—dark feather inside of Waterhen's wing; body,—dark red brown silk, black hackle for legs—tail two strands of the same. 2. Olive Blue,—feather of Starling's wing, body light olive silk, and red hackle. 3. Red Clock,—wings and legs red; Peacock's brown herl, and bright red silk for body. 4. Little Brown,—feather from inside of Woodcock's wing, red copper coloured silk for body, and brown hackle for legs. 5. Blue Midge,—feather of Waterhen's neck,—lead coloured silk for body, grizzled hackle for legs. 6. Great Brown,—feather from the hen Pheasant's wing,—dark orange silk for body, brown red hackle for legs,—tail do. 7. Yellow Legged Blue,—feather from inside of Teal's wing, or lightest part of Starling's wing,—straw coloured silk for body, legs yellow hackle,—tail do.

FOR APRIL.

1. Dark Blue,—same as March. 2. Dun Midge,—lightest part of a Thrush's quill feather,—pale yellow silk, ribbed with light orange,—legs yellow hackle. 3. Spider Legs,—rusty coloured feather from Feldfare's back,—lead coloured silk for body, grizzled hackle for legs. 4. Sand Fly,—ruddy mottled feather of hen Pheasant's wing,—reddish fur from Hare's neck, ribbed with light brown silk,—ginger coloured hackle for legs. 5. Green Tail or Grannam,—wings inside of hen Pheasant's wing,—body lead coloured silk, with Peacock's green herl for tail,—legs ginger hackle. 6. Inside of Woodcock's wing,—body orange coloured silk neatly ribbed,—hackle from a grouse for legs.

FOR MAY.

1. Grey Midge,—feather from Woodcock's breast,—body of pale yellow silk. 2. Yellow Sally,—pale yellow feather,—body yellow silk,—legs yellow hackle. 3. May Brown,—ruddy grey,—feather from Partridge's back,—olive coloured silk ribbed with light brown for body,—legs, hackles of an olive colour, tail do. 4. Pale Blue,—Sea Swallow for wings,—yellow pale silk for body, ribbed with sky blue,—pale yellow hackles for legs,—tail do.,—Little Stone Blue,—feather from Blackbird inside the wing, or Swift,—brown silk for

body, brown hackle for legs. Stone Fly,—Mallard's feather from the back,—very large for wings,—two strands of yellow, and one of drab,—Ostrich herl neatly ribbed,—tie with brown silk.—horns and tail, black cat's whiskers.

FLIES FOR JUNE.

1. Little Dark Blue,—inside of Waterhen's wing,—lead coloured silk for body, legs yellowish dun hackle, tail Rabbit's whisker. 2. Pale Blue—light part of Starling's quill feather for wings, pale yellow silk for body, pale yellow dun hackle for legs and tail. 3. June Dun—a feather from Dottrel's back, hackled on a body of blue Rabbit's fur and drab silk, dun hackle for legs. 4. Twitchbell—inside of lightest part of Starling's quill feather for wings, brown hackle for legs, brown Peacock's herl for body.

FLIES FOR JULY.

1. Little Olive Blue—Feather of Starling's wing dyed in onion peelings, lead coloured silk for body, ribbed with yellow, dun hackle for legs, stained like the wings, Rabbit's whiskers for tail. 2. Little Black Ant—feather of a Bluecap's tail for wings, black Ostrich herl dressed small in the middle for body, brown hackle for legs. 3. Red Ant—Lark's Quill feather for wings, cock Pheasant's herl from tail for body, red hackle for legs. 4. Little Blue—Bullfinch's tail feather for wings, dark blue silk for body, dark blue hackle for legs, tail do.

FLIES FOR AUGUST.

1. August Brown—feather from hen Pheasant's wing,—fern coloured fur from Hare's neck, ribbed with pale yellow silk,—grizzled hackle for legs,—tail do. 2. Light Blue,—inside of Snipe's wing,— body light Drab silk,—tail and legs grizzled hackle. 3. Cinnamon Fly,—feather from Landrail,—orange and straw coloured silk for body,—ginger hackle for legs. 4. Light Blue,—inside of Snipe's wing,—light drab silk for body,—legs and tail grizzled hackle. 5. Dark Blue,—feather from Waterhen inside the wing,—reddish brown silk for body,—legs and tail brown hackle. 6. Orange Stinger,—taken from middle of August to the end of September—feather from Starling's quill,—the head brown—the tail orange silk,—for body and legs, furnace hackle. 7. Grey Grannam,—dark feather

from night Hawk or brown Owl,—red Squirrel's fur and fawn coloured silk for body,—ginger hackle for legs.

FLIES FOR SEPTEMBER.

1. Light Olive Blue,—Dottrel's wing,—body pale white French silk,—legs and tail pale blue hackle. 2. Dark Olive Blue,—wings inside of Waterhen's wing,—body lead coloured silk,—black hackle for legs,—tail Hare's whiskers. 3. Small Willow Fly,—wings inside of Woodcock's wing feather,—body mole's fur and yellow silk,— brown hackle for legs.

OCTOBER.

1. House Fly,—lark's quill feather,—light brown silk,—ribbed with dark Ostrich herl for body,—legs grizzled hackle. 2. Small Olive Blue,—wings Starling's feather stained with onion peelings,— yellow silk for body,—legs olive stained hackle. 3. Dark Grey Midge,—wings dark grey feather of a Partridge,—body brown silk,—legs grey Partridge hackle.

RED PALMER.

Body greenish herl of Peacock,—ribbed with gold tinsel,—wrapt with red silk,—red hackle over all.

BLACK PALMER.

Body dark Peacock's herl,—ribbed with gold tinsel,—green silk, black, brown or red hackle over all.

MAY FLIES.

THE YELLOW, GREY, AND GREEN DRAKES.

These flies, which are known as May flies, afford great sport. Trout and Greyling are so partial to them that they refuse all others during the time they are on the water, but they are not common to all rivers. The Driffield, Derwent and other Yorkshire streams, have them in great abundance. The best chance with the artificial May fly, is when there is wind stirring sufficient to cause a pretty considerable curl on the water. The *Yellow Drake* may be made in this way,—a Mallard's back feather dyed yellow; for wings, Cock's hackle dyed yellow; underneath the wings to make them stand upright, yellow camlet, ribbed with brown silk for body; tail, two hairs from Squirrel's tail. *Grey Drake,*—wings from Mallard's back feather, black Cock's hackle underneath; body sky blue camlet ribbed with copper coloured Peacock's herl; tail from Squirrel. *Green Drake,*—same as yellow except the wings, which must be from a Mallard's feather dyed a yellowish green.

I have not deemed it requisite to introduce any illustrations of flies, because I cannot conceive that any really beneficial results are obtainable by merely showing the difference on paper between natural and artificial flies. Catch the natural fly, imitate it as closely as possible; put your made fly into a tumbler of clear water, then if the size and the prevailing colours as to body and wings resemble your copy, you are all right. This appears to me the best comparative illustration.

I beg to suggest to those who have opportunity and leisure, that they might at the cost of a little trouble, make a collection of all the flies that come on the waters, where they are accustomed to angle. They are easily caught and preserved, and if classed according to the months during which they were found, would be useful and interesting to themselves and friends, if only to refer to when manufacturing flies.

HOW TO MAKE A HACKLE FLY.

Take a hook of the required size, between the finger and thumb of your left hand, with the point towards the end of your finger, place the gut along the top of the shank, and with the silk bind them tightly together, beginning half way down the shank, and wrap the end, take two turns back again which will form the head of the fly; lay the feather along the hook, the point towards your left hand, and take three turns over it with the silk, clip off the points of the feather, and bind it neatly round till the fibre is consumed, bring the silk round the root of the feather to bind to the end of the tail of the fly. Cut off all superfluities and fasten off by the drawn knots, then with a needle trim the fibres and your fly is made.

TO MAKE A WINGED FLY.

Have your materials ready, wings silk &c., of the colour you require, then take a hook between the forefinger and thumb of your left hand, with the point towards your forefinger, place the gut at the top of the shank, and with the silk bind them tightly together, bind all tight within two or three turns of the shank of the hook. Take the feather for wings, lay the feather's point the proper length between your finger and thumb along the hook, and take two or three turns over it for the head of the fly, and bind the gut between the second and third fingers of your left hand, and with the scissors clip off the root end of the feather, wrap the silk back again once under the wings, setting them upwards; with the point of the needle divide equally the wings crossing the silk between them. Lay the hackle for legs, root end towards the bend of the hook, wrap your silk over it and so make the body of your fly, then take the fibre end of the hackle, rib the body of the fly neatly with it, till you reach the silk hanging down, wind the silk twice or thrice over the hackle, fasten with the usual knots, and your fly is complete.

MATERIALS REQUIRED FOR MAKING WINGED AND HACKLE FLIES.

In the manufacture of winged flies a great variety of feathers are required. Procure those of a Mallard, Teal, Partridge, especially the tail feathers; also, the wings of a Starling, Jay, Landrail, Waterhen, Blackbird, Fieldfare, Pheasant Hen, Pewitt's Topping, Peacock's Herl, green and copper coloured, black Ostrich herl, Snipe, Dottrel, Woodcock and golden Plover's wings, the tail feathers of the blue and brown Titmouse, and also Heron's plumes. Dubbing is to be had from old Turkey Carpet, Hare ears, Water Rat's fur, Squirrel, Mohair, old hair cast from young cattle, of a red, blue, brown, black and fawn colour from behind a Spaniel's ears, and from the fur of a Mouse, and note, Martin's fur is the best yellow that can be had. In regard to Silks be careful to suit the colour of the silk (at least as much so as you possibly can,) to the hackle you select for dubbing with. Thus with a Dun hackle, use yellow silk; a black hackle, sky blue; a brown or red hackle, red or dark orange do.

The above selection of silks and dubbing are for Palmers and winged flies generally. It is a good plan however to take and wet your dubbing previous to making use of it, because when dry it may appear the exact colour you need, yet wetted quite the reverse. To acquire an accurate knowledge of any dubbing, hold it between the sun and your eyes. Mohairs may be had of all colours, black, blue, yellow and tawny, from *feuille morte* a dead leaf, and Isabella which is a whitish yellow soiled buff.

TO MAKE A PALMER FLY.

Take a length of fine round silk worm gut, half a yard of silk well waxed, (wax if possible of the same colour,) take a No. three or four hook, hold it by the bend between the forefinger and thumb of the left hand, with the shank towards your right hand, and with the point and beard of your hook not under your fingers, but nearly parallel with the tips of them, then take the silk and hold it about the middle of it with your hook, one part laying along the inside of it to your left hand, the other to your right; then take that part of the silk which lies towards your right hand between the forefinger and thumb of that hand, and holding that part towards your left tight along the inside of the hook, whip that to the right three or four times round the shank of the hook towards the right hand, after which take the gut and lay either of its ends along the inside of the shank of the hook, till it comes near the bend of it, then hold the hook, silk and gut tight between the forefinger and thumb of your left hand, and afterwards put that portion of silk into your right, giving three or four more whips over both gut and hook, until it approaches the end of the shank, then make a loop and fasten it tight, then whip it neatly again over both silk, gut, and hook, until it comes near to the end of the hook, make another loop and fasten it again; now wax the longest end of the silk again, then hold your Ostrich strand, dubbing on whatever you have selected, and hook as at first with the silk just waxed anew, whip them three or four times round at the bend of the hook, making them tight by a loop as before, then the strands to your right hand and twisting them and the silk together with the forefinger and thumb of the right hand, wind them round the shank tight, till you come to the place where you fastened, then loop and fasten again, then take your scissors and cut the body of the Palmer into an oval shape, that is, small at the head and the end of the shank, but full in the centre; don't cut too much of the dubbing off. Now both ends of the silk are separated, one at the bend, the other at the end of the shank, wax them afresh, then take the hackle, hold the small end of it between the forefinger and thumb of your left hand, and stroke the fibres of it with those of your right, the contrary way to what they are formed; hold your hook as at the beginning, and place the point of the hackle on its bend with that side growing next to the Cock's neck upwards, then whip it tight to the hook, but in fastening, avoid if possible, tying the fibres; the hackle now being fast, take it by the large end and keeping that side which lies to the neck of the Cock to the left hand, begin with your right hand to wind it up the shank upon the dubbing, stopping every second turn, and holding what you have wound tight with the fingers of your left hand, whilst with a needle you pick out the fibres unavoidably left in; proceed in this manner till you come to where you first fastened, and where an end of the silk remains; then clip off the fibres of the hackle which you hold between your finger and thumb close to the stem, and hold the stem close to the hook, afterwards take the silk in your right hand and whip the stem fast to the hook, and make it tight: clip off the remaining silk at both bend and shank of the hook, and also all fibres that start or don't stand well, and then your fly is complete.

GOLDEN PALMER.

Take the hair of a black Spaniel for dubbing, ribbed with gold twist, and a red hackle over all.

SILVER HACKLE PALMER.

The same dubbing as for the Golden Palmer, silver twist over that, and a brown red hackle, and note, when you make Golden or Silver Palmers, and when whipping the end of the hackle to the head of the hook, do the same to the twist whether Gold or Silver, first winding on the dubbing, observing that they lie flat on it, then fasten off and proceed with the hackle, or you may wind the hackle on the dubbing first, and rib the body with either of the twists afterwards. Palmers may be made so as to suit all waters by making them of various colours and sizes, and it is a good plan to fish with a Palmer until you know to a certainty what fly is on the water. Hackles for Palmers should consist of red, dun, yellow, orange and black, they should not by any means exceed half an inch in length. A strong brown red hackle is exceedingly valuable. Any person who can make a Palmer will make winged flies without difficulty.

TO MAKE HACKLE FLIES.

Select a feather the colour you want, and whose fibres are of the length suitable for the size of the fly you wish to dress. Strip off all superfluous fibres, leaving on the stem of the feather no more than you require for your fly, then having previously waxed about half a yard of fine silk of whatever colour you deem best, take your gut or hair and hook into your left hand, lay the gut inside the shank of the hook nearly down to the bend, then whip the gut and hook, at the end of your hook together, then lay your feather the reverse way from the top of the feather on to the gut and hook, make fast the feather with your silk, then wind your silk on the hook as far as you intend the fibres to extend, holding the hook, gut and silk in your left, with your right wind the fibres down to the silk and make all fast, then wind the remaining part of the gut and hook as far as nearly the bend of the hook with your silk, and fasten; wind your silk back again to the feather, make all fast, cut off the remains of the silk, smooth down the fibres, press them between your finger and thumb, and having arranged them to your mind, the fly is completed. Instead of carrying the silk back again to the feather from the bend of the hook, you may finish there, if you prefer doing so. I prefer the former. Making hackle flies is such an easy matter, that any person with any ingenuity and attention, may soon become a proficient in fabrication of them, and by diligent observation as to the size, colour, and peculiarities of the great variety of natural flies, which make their appearance on the water at particular seasons and hours of the day, he will at all times be enabled to pursue his diversion with the best chance of success. Nature best followed best secures the sport.

WORM OR BOTTOM FISHING.

You may take Trout in February with the worm if the weather is mild, and continue to do so until the end of October. It is a most alluring and destructive bait, and requires more skill to fish it properly than is generally supposed. After rain, when rivers or brooks are somewhat beyond their usual bounds, a well scoured lob worm will take the best of fish. For worm fishing you must have a yard of good gut attached to your cast line, which line ought to be of the same thickness from the gut to the loop of your reel line, your hooks may be a trifle larger in the Spring than in the Summer, and should be tied on to the gut with good strong red silk; two No. 4 or 5 shot corns, partially split, and then fastened upon the gut about five or six inches from the hooks, and from two to three from each other, are generally sufficient in a strong water to sink your worm to the requisite depth, but in low and fine waters, use two of No. 6, and sometimes one will be sufficient. In worm fishing never attempt to fish down, but always up a stream, and when you are aware that you have a bite, slacken your line a little in order to give time to the fish to gorge, then strike quickly, but not too hard, and land your prize without delay; you need not make more than two or three casts in one place, because if there is a fish he will in those casts either take or refuse your bait. In summer when the water is low and fine, and the thermometer about seventy-five Farenheit, capital sport may be had with well scoured Brandlings, perhaps this sort of fishing is *nulli secundus*, inferior to none in the exercise of skill and ingenuity. The immortal Shakespeare, must surely have fished the worm in clear waters, for he says, "the finest angling 'tis of all to see the fish with his golden fins, cleave the golden flood, and greedily devour the treacherous hook." In the Spring you must give your fish more time before you strike them than in the Summer; because having been sickly and altogether out of order, and not yet having recovered his usual strength and activity, he bites but languidly, and does not gorge so quickly as when in prime condition. When you find Trout pulling or snatching at the worm, which may be termed runaway bites, and when in fact they neither take it nor let it alone, it is a sign they are full, and the best plan to effect a capture under such circumstance is to strike that moment they touch your bait, for if you do not succeed by a snap, but allow them time, they will only play with it for a few moments, and then finally leave you in the lurch. In concluding my observations on worm fishing, I can with confidence affirm that it is, as a bait for Trout, the most destructive and certain agent the angler (taking the season through) can make use of. The author of Don Juan certainly did not flatter a worm fisher, one part of his assertion however is undoubtedly true, the worm was at one end, but it did not necessarily follow, that a fool was at the other. His poetic and satirical lordship probably never saw Trout taken with the worm in a clear stream, if he had I think he would have been satisfied that there was nothing foolish about it. Osbaldiston in his *British Field Sports*, under the head of *Allurements for Fish*, recommends the gum of ivy, he says, "take gum ivy and put a good deal of it into a box made of oak, and rub the inside of it with this gum; when you angle, put three or four worms into it, but they must not remain long, for if they do, it will kill them, then take them and fish with them, putting more into the worm-bag as you want them. Gum ivy flows from the ivy tree when injured by driving nails into it, wriggling them about and letting them remain for some time; about Michaelmas is the best time to procure it. Gum ivy is of a red colour, of a strong scent, and sharp pungent taste." When fish are disposed to feed, you need not use gum ivy; the attractions of a bright and clear scoured worm are quite sufficient without any such adjunct.

TROLLING WITH THE MINNOW.

You must for this kind of Angling, have a tolerably strong Rod and tackle, you may begin trolling about the middle of March, and continue to the end of October. The very best of fish are taken with the Minnow, it is an active bait to fish with, and keeps the Angler pretty well on the *qui vive*. When the water is in order, that is, after it is a little swollen and discoloured by recent rain, it frequently proves a most destructive bait, and will take Salmon as well as Trout. Those Anglers who are desirous of a few good fish, will find it their interest to use it on every suitable occasion, independent of the good fish to be had with it, it is next to fly fishing, the most animating and exciting method of angling. To make your Minnow spin well, one or two swivels should be used, attached to the gut, which should be about a yard in length and of fine and good quality. In fishing the Natural Minnow with two hooks, one of them must be large enough to pass through the body of the bait, going in at the mouth, and passing out at the tail; the other, rather larger than a May-fly hook, should go through the under, and pass out at the upper lip. In trolling with only two hooks, be careful to give your fish time to gorge, otherwise by striking too quickly, you will miss your prize by pulling the bait out of his mouth. With three or more hooks, which is termed fishing at snap, you cannot strike too soon as the fish is generally caught by one of the loose hooks. If the fish you have hooked be not too heavy, the best plan is to land him at once by a quick and sudden jerk. In fishing the Minnow, if in still, deep water, let it sink a little at first, then draw it quickly towards you, making the bait spin well and briskly, which is effected by the swivel. In streams, especially if they be rapid, cast up and down, but chiefly athwart, by so doing your bait shows greatly to advantage. Trolling in the Tees is not much practised; the difficulty of procuring Minnows at the precise time when wanted, is I suppose the reason. But there are artificial Minnows which in heavy waters will kill well; those sold by Frederick Allies, South Parade, Worcester, and by Farlow, Tackle Maker, in the Strand, London, are excellent, the price for Trout reasonable, two shillings and six pence. The former is styled the Archimedean, the latter the Phantom Minnow, which collapses when struck by a fish. The best river I have ever trolled in, and I do not suppose there is a better in England, is the

Eden, which takes its rise a few miles from Kirby Stephen, in Westmoreland, thence to Carlisle, and so seaward, running for the most part over a gravelly and sandy bottom, and full of good Trout, so that splendid sport may be had by trolling when the water is in proper order. The Greta is an excellent trolling stream, but the fish are not near the average weight of those in the Eden. It is not a bad plan when the water is low and fine, and Minnows are easily procured, because you may then see where they are, especially on a sunny day, to catch as many as you want, (which you may do, with small hooks baited with very small red worms,) and then cure them. Of course those cured are not so good and durable as the fresh, but still they are found to take fish very well. And thus provided with artificial and pickled auxiliaries, the indefatigable troller will never be brought to a stand. For what can be more provokingly annoying to an angler, than to have to leave off in the very midst of sport, merely for want of baits?

MAGGOTS

May easily be had; any description of flesh exposed to the sun is soon full of them, for choice I should prefer horse flesh; when sufficiently large they are an excellent bait for Trout; preserve them in tin case (with holes to admit air,) filled with bran, where they will scour a trifle and keep alive some days; when you fish with them, use a Palmer sized hook, and a single No. 5 shot corn, and when the water is as low or almost as much so as it well can be, your gut need not be leaded at all.

WASPS AND HUMBLE BEE GENTLES.

These Gentles are excellent for both Trout and Chub, preserve them the same way as Maggots, and use the same sized hook.

DOCKEN GRUB.

This grub is found, as it name indicates, at the roots of dockens: the body of it is somewhat similar to a Maggot, it is a good bait for Trout and Chub, and may be kept some time in a woollen bag containing fine sand; fish will often take it when they refuse the worm, you may begin to fish with it in February and continue to do so during the season. Small May-fly hook and one No. 5 shot corn.

CREEPER,

Found underneath stones having a little water and gravel or sand underneath them, may be kept in a May-fly horn, but soon die for want of water; a good bait early and late, or in streams on a hot day. A No. 6 shot corn and May-fly hook, fished like the worm.

CADISS, OR CAD BAIT,

Found in brooks or rivers, encased in little straw or gravel husks: a curious little grub similar to a gentle in size, with a dirty yellow body and black head. Palmer sized hook, shot corn No. 6, or your hook slightly leaded on the upper part of the shank, round which have the hackle of a Landrail or dyed Mallard. Kills well with hackle when the water is slightly discoloured.

WORMS.

"You must not every Worm promiscuous use."

Gay.

The best for Spring fishing are the Marl or Meadow worms, the Gilt Tail, the Squirrel Tail and the Brandling, are excellent in Summer. A Lob Worm well scoured is a good bait early in the morning, either in Spring or Summer. When you fish with the Brandling, it is a killing way to have two on your hook, letting the head of the second Brandling hang a little way over the tail of the first, or you may put heads and tails together; always procure your worms, and put them to some good moss, some time before you want them; after three or four days, by adopting this method, they will be clean, bright and tough; a glazed earthen jar is the best thing to keep them in, and in Summer set your jar in as cool a place as possible; by attention in changing your moss every fourth day, or so, you may preserve and keep your worms a long time. Moss from heaths and waste lands, is the best you can get; always be careful to pick from the moss all blades of grass, leaves, or dirt adhering thereto. Put your worms into water if you want them scoured quickly, and let them remain in it for twenty minutes or half an hour, they come out in an exhausted state, but soon recover on being put into good clean moss. Bole Armoniac will also scour them very speedily. As to gum ivy and ointment put to worms to entice fish, such practises I hold to be mere matters of fancy, and I do not deem it necessary to give instructions in reference thereto. It is my opinion only time and trouble thrown away, and you may depend upon this as a fact, that if fish will not take a bright clean worm, the addition of unguents will be found useless. As I have observed elsewhere, it is the eye and not the sense of smell (if they have any) which guides, influences, and directs fish in their choice of food.

You may breed worms in abundance by the aid of decayed vegetables and leaves, mixed with marl or any kind of soil; the Brandling or Red Worm are found in Pig's and other dung, also in Tanner's bark.

SALMON ROE.

Salmon Roe is such a destructive bait for nearly all kinds of fish, and Trout in particular, that I know nothing comparable to it. It is moreover a bait requiring but little skill in the use of it. After a flood, and before the water clears, is the best time for fishing with Roe. Log, or still water having a gravelly, or sandy bottom, is the place to be selected, and you may use three or four stiff rods, placed at convenient distances from each other. You can also have floats if you like, by doing so you will immediately perceive when you have a bite. It is a good plan previous to casting in your lines, to sound the depth of the water, which you may do easily enough with a string leaded for the purpose; because, it is of material consequence that your Roe should lie at, or very near the bottom of the water. A hook about the size of a Limerick May fly hook, is quite large enough to put your roe on, which should be in regard to size about that of a French Bean or marrow fat Pea.

Salmon Roe is cured and preserved by spreading it upon thin layers of cotton wool, pack the layers on each other and cover them tightly up, so as to exclude air; glazed jars covered with bladder over the tops of them are the best to keep your Roe in. When you want to use it, mix the Roe with a little wheaten flour and gum water, to cause adhesion to the hook. In concluding this notice of Roe, I cannot refrain from expressing a hope that gentlemen will abstain from the use of it. By the purchase of Roe they hold out a premium to Salmon poachers who annually destroy immense numbers of spawning fish solely for the sake of the Roe, the high price which it commands encourages them in their illegal pursuits. If there were no buyers of Roe there would soon be a visible increase of Salmon.

DYING FEATHERS FOR FLY MAKING.

For dying feathers use clear soft water; to strike the colour add to each pint of water a piece of alum about the size of a walnut; to dye white feathers yellow, boil them in onion peelings or saffron. Blue feathers by being boiled as above become a fine olive colour. To dye white feathers blue, boil them in Indigo, by mixing the blue and yellow together, and boiling feathers in the mixed liquid, they become green. Logwood dyes lilac, or pink; to turn red hackles brown, boil them in copperas. To stain hair or gut for a dun colour, boil walnut leaves and a small quantity of soot in a quart of water for half an hour, steep the gut till it turns the colour you require. To stain gut or hair blue, warm some ink, in which steep for a few minutes, then wash in clean water immediately; by steeping hair or gut in the union dye, it will turn a yellowish green, and in gin and ink it becomes a curious water colour.

TO MAKE STRONG WHITE WAX.

To make strong white wax, take three parts of white rosin and one of mutton suet; let them simmer ten minutes or so over a slow fire, dropping in a small quantity of essence of lemon, pour the whole into a basin of clear cold water, work the wax through the fingers, rolling it up, and then drawing out until it is tough. It cannot be worked too much. By using this wax the pristine colours of the silk you use in fly making are preserved; common shoemakers wax soils the silk too much.

FISHING PANNIERS OR BASKETS.

The French Baskets are the neatest, lightest and most durable, being closely woven, they very much exclude the air, so that fish look better on being taken out of a pannier of that description; many of the English made fishing baskets, are only of clumsy construction, and have the fault of being too open in the weaving, admitting far too much air, whereby, particularly on windy days, your fish become dry and shrivelled.

LANDING NETS.

Landing nets round or square, are made of strong silk or best water twine cord. Those nets having a joint in the centre of the shank, are most convenient when travelling. It is not advisable to have too deep a net, as your flies become very often entangled in such a one, and cause much trouble and loss of time in extricating them; therefore a net that is sufficiently deep to hold a good fish without admitting a possibility of escape, is the kind of net you require.

WINCHES OR REELS.

Winches may be bought at all tackle shops, and of any size you wish. My remarks on them extend only to this, that they are very useful appendages to any rod, and give you great advantage over a good fish, enabling you to give line and play him as you like; should a breakage of your top or other part of your rod happen, you have it safe, being held by your reel line. A light winch that will hold from 25 to 35 yards of line is sufficient for Trout. A Salmon winch should be capable of holding from 50 to 80 yards of line.

GUT AND HAIR.

In selecting gut for Trout fishing, choose that which is round and fine. What is termed manufactured gut, may be had at most tackle shops, it is exceedingly fine but not durable, the best I ever met with was at Rowel's, at Carlisle, 1d. per length. Hair should be bright, round and strong, chestnut hair suits moss or discoloured waters, if you can procure hair of a light or bluish tint, that is the best of colours; both gut and hair should be wet when knotted.

RODS.

The three distinguishing characteristics of a really good fly rod are strength, elasticity, and lightness, such rods are to be bought in the London tackle shops for a pound; these rods are perfect as three or four piece rods, but I much prefer one for my own use in only two pieces, such a rod is more readily put to, and taken from together than one consisting of three or more joints; not so liable to get out of order, and has a truer bend with it when subjected to pressure. I recommend a rod having a root 9 feet, and a top of 5 feet, making together 14 feet in length, as the most useful; a fir root, and top of good sound lance wood, well painted, ringed and varnished, makes a neat and serviceable rod. For trolling, your top should be stiff and strong. For worm not so pliable as your fly top.

LINES.

Lines composed entirely of hair, are lighter on the water than those made of silk and hair mixed, perhaps the latter is the stronger line of the two, but it both carries more water and is more expensive. A winch line should be for Trout from 25 to 35 yards in length, and may be bought at all tackle shops, at the rate of a 1d., 1½d. and 2d. per yard, according to quality; at so cheap a rate, it is scarcely worth while to make your own line, which you may do by the purchase of a little machine for twisting, or you may use goose quills, which is however but a slow and tedious process.

HOOKS.

The best hooks are Kendal, Limerick, and Carlisle; I prefer the Limerick for fishing the natural flies, they are all however very good. Some anglers are partial to the Kirby bend, but perhaps you get better hold of your fish with the sneck bend hooks. If you purchase wholesale, you get 120 hooks for a shilling, if by retail at tackle shops, generally 6 a penny, or 72 for a shilling; so that wholesale you have about 50 more hooks for your money.

REMARKS ON FISHING GARMENTS.

With Cordings, Fishing Boots, and Macintosh Coat, you are weather proof; neither the water from above or below can affect you; by the aid of the boots you keep your feet perfectly dry, the coat enables you to continue fishing during the heaviest showers, and in Summer especially, when the flies and insects are beat down by such showers, the best of fish are then on the move; without the India Rubber Garment, you may get thoroughly wet in ten minutes. If you find shelter you probably loose some good sport, and if not, by continuing your fishing, you become so cold, wet, and exceedingly uncomfortable, that you generally deem it adviseable to proceed home with as little delay as possible. When the day is fine, and the water repeller not needed, avoid light, or glaring colours; brown, green, or grey garments are most suitable, particularly when the water is low and clear.

HEALTH,—CAUTION.

If your feet are wet either in Spring or Summer, do not, if you regard your health, sit down above two or three minutes. You may frequently have occasion to wait some considerable time by the water side, looking out for the expected feed, and consequent rising of the fish; at such times keep walking about in preference to sitting, which is the best way to avoid catching cold. When you return home loose not a moment in changing your wet garments. Colds and Rheumatism are the pains and penalties anglers are liable to, who do not follow the above advice.

THE EYE, THE ONLY ACUTE FACULTY IN FISH.

Trout, however quick sighted they may be, are like all the finny tribe, supposed to be incapable of hearing, in consequence of the density of the element in which they exist. Water has long ago been proved to be a non-conductor of sound, and if fish are possessed of any faculty of the kind, it must be the dullest imaginable. From the horny construction of the palate, their taste cannot be acute, and their sense of smelling (judging from the medium by which all odours are conveyed to them,) must be peculiarly defective. Taking the above suppositions to be correct, it is of course clearly apparent that they must be guided solely by the eye in the selection of their food; for instance, when fish are stupefied or fuddled as it is termed, I do not suppose their olfactory organs are affected by the berry or drugs, used to intoxicate or kill them. I am persuaded, that small balls of paste or bread would, if offered to them at the same time, be devoured at precisely the same rate as those prepared with unguents or drugs.

The formation of fish is peculiarly adapted to water, through which they glide with the greatest facility; their motions being regulated by the fins and tail; the tail indeed being to the fish precisely what a rudder is to a ship. The air bladder in fish is another wise provision of nature, by means of it they can remain for a long time under water; still they must from time to time take in supplies, for if during a severe frost the ice be not broken on ponds, the fish therein would perish for want of air. Some fish are much more tenacious of life than others; Roach, Perch and Tench, have been conveyed alive, for stocking ponds, thirty miles, packed only in wet leaves or grass. One thing is quite certain as regards all fish, viz., that they live longer out of their natural element in cold than in hot weather. A clever invention for the transport of fish has come under my notice; an account of this machine may prove interesting to some persons, and therefore I insert it.

THE TRANSPORT OF TROUT AND GREYLING.

The Apparatus consists of a tin case, separated into two parts by an open work partition. In one of these the fish are placed, and in the other is fixed a mechanical contrivance for keeping up a considerable supply of air in the water.

In November, 1853, 33 Greylings were sent from the Wye at Rowley to the Clyde at Abington, a distance of about 250 miles with the loss of only two fish.

The Apparatus is composed of a zinc cylinder, about three feet high and two feet in diameter, with a strong iron handle running round the middle; to the top, a small force pump is attached, and by this fresh air is forced through a star shaped distributor at the bottom of the cylinder; a ring to bring the fish up for inspection, and a loose concave rim to prevent splashing over, complete it. A drawing with particulars was deposited with the Society of Arts, in London.

THE NATURAL ENEMIES OF FISH.

Fish have so many enemies that were it not for the millions of embryo or spawn deposited by the female, the breed of Salmon and Trout (to say nothing of other species) would long since have become extinct. Eels, fish, birds, water rats, toads, frogs, and last but not least, the water beetle,[8] prey upon the ova, spawn and young fry; floods also sweep away and leave on banks, or rocks, a considerable quantity of spawn, which of course comes to nothing. Escaping the above perils and causalities, and arrived at maturity, they become the prey and food of the otter and heron, king's fisher, gull, &c., who emulate man in their destructive propensities. The larger fish also prey upon the smaller. Luckily otters are not so numerous in any English river as they used to be. Night lines, shackle, rake and flood nets, and other devices not at all creditable to those who use them, and to which I shall not further allude, make terrible havoc amongst fish, and mar and spoil the fair and honest angler's sport, but in most rivers and brooks of Trouting celebrity, such practices are greatly on the wane. Proprietors will not sanction such wholesale destruction; and now almost universally adopts measures for the detection and punishment of such depredators.

LAWS RELATIVE TO ANGLING.

It would occupy too much space to be diffuse in reference to angling laws; I shall therefore briefly observe that all persons discovered robbing fish ponds during the night, and all persons found poisoning fish are liable to transportation; all persons using nets, listers, snares or other unlawful devices, are liable to the forfeiture of such nets, &c., and also subject to a fine at the discretion of the magistrates before whom such offenders may be brought; and also, that any person angling in any brook or river without the permission of the proprietor or proprietors of such river or brook, is liable to a penalty as a trespasser, and also to the forfeiture of any fish he may have caught.

OBSERVATIONS IN REFERENCE TO THE EFFECT OF THE WEATHER ON FISH.

Your sport in angling, whether top or bottom, materially depends upon the state of the atmosphere. He who has paid some attention to the effects of weather on fish, knows pretty accurately the extent of the sport to be looked for, when the wind is in particular arts. An East or N. East wind shuts out all hope of diversion, whilst a Southerly or South West wind, is the wind of all winds for the angler. However, as fish must feed at some time, let the wind be as it will, an angler who is particularly in want of a few Trout, may succeed in obtaining small ones with the fly in an East or N. East wind, provided the wind has been in that quarter some days, and there is feed on the water. Any sudden change in the wind affects the fish, and they will sometimes give over, or begin to feed, on such changes taking place, just as it happens to veer into the wrong or right quarter. After white frosts in the Spring of the year, you need not expect much, if any sport. Frosty nights with bright sunny days following, accompanied with East or N. East winds, are precisely those sort of days, when a man had better refrain altogether from attempting to take fish with the fly, or with any kind of bait. During the Summer months, the colder the wind blows, the better sport you will have with the artificial fly. On cold stormy days in Spring, with wind West or N. West, accompanied with heavy snow, rain, or hail showers, good fish are usually roving about, and then your sport is of the best. Either in Spring, or Summer,

"With a Southerly wind, and a cloudy sky,

The angler may venture his luck to try."

WHAT CONSTITUTES A GOOD FISHING DAY.

It is of the greatest consequence to acquire a correct estimate of what really constitutes a good fishing day; and not put too much faith in the advice of the author who wrote an article on angling, which is published in *Brewster's Encyclopædia,* who tells us to follow the example of the navigator, who does not wait for a favourable wind, but goes to sea at once, to seek for one; not to sit at home on the look out, but go to the river in all weathers. The three great essentials of a good Trouting day, are water, wind, and cloud, if there is a failure in all three, you are better at home, at least that is my humble opinion. If a deficiency or partially so in any, expect only moderate sport, but if all three are in unison, then you may fairly calculate on excellent diversion. There is nothing like a South West wind for holding forth a promise of a cloudy day. As to the water, the second day after a heavy fall of rain is often the best. The wind however sometimes (too frequently indeed) veers into the North West, or further on that day, and if the barometer rapidly rises at the same time, there will be too much sunshine; on the third, if the wind veers to the South West, the day will probably be too dark; for a dull day occurring about new and full moon, is seldom a good angling day. A man whose avocations do not permit him to angle in all weathers, will therefore do well to select a day, when the three great essentials of his sport, wind, water, and cloud, are in his favour.

Note.—An angler is so dependent on the weather that he should omit no opportunity of acquiring meteorological knowledge. Electric influences guide and coerce fish in a wonderful manner.

ON EARLY RISING IN CONNECTION WITH ANGLING PUR-
SUITS.

Thousands of the dwellers in "the modern Babylon," and indeed in all large cities and towns, never saw the splendour of a rising sun. Tens of thousands never heard the sylvan choristers performing their morning's concert, filling with their melody, nature's own, the woods and groves wherein these feathered songsters "sport, live, and have their being." Whilst millions of men are sunk in the arms of "the drowsy god." What is the angler about, has he slept soundly, and then awoke in the very nick of time? Or have his slumbers been somewhat broken and disturbed by dreams of crafty old Trout? No matter, he is astir, he has pocketed his tackle, and not neglected something for the inner man; rod and net in hand, he is off and away frequently before, but seldom later, than the rising lark proclaims with joyous notes the coming day; full well, he knows the advantages of an early move during the Summer months; the morning is all in all, the best part of the day to him; so, buoyant with hope he progresses at an easy rate towards the scene of his triumphs, or disappointments, as the case may be. An angler of early habits during the Summer months sees a great deal of animated nature, and ought to know as much of the habits of birds, animals, insects, &c., as any man. At early morn the great volume of nature lies open for his inspection, if he be intelligent and curious, he will soon become a naturalist, whether his path leads through the woods, the lowlands, or over the uplands, he is pretty sure to meet with something to gratify, instruct and amuse. Independent of the varied attractions of nature, the early rising angler always has the best Summer sport. Large fish invariably feed more freely in the morning than during any other portion of the day, evenings occasionally excepted; he also avoids the greater heat by getting home a.m., indeed after twelve o'clock on a Summer's day your shadow falls more or less upon the water, and scares the fish. Independent of that, they usually cease to feed by that time.

In streams where piscatorial rights are cherished, and protected to their fullest extent, Trout are frequently found to be much smaller, than might naturally be supposed; the fact is, that in good breeding waters strictly preserved, Trout soon become so numerous that the supply of food is inadequate to their wants; a state of things which in rural parlance is termed, as having more stock than the pasture will carry; a numerical reduction, to some extent in such streams is therefore extremely beneficial. Better fish are sometimes met with in free waters than in preserves, solely because they have had abundance, and variety of food. In all moor becks, plenty of small Trout are found; such waters are excellent for breeding, but as very little nutriment comes from peat or waste lands, they are generally dwarfish in size, and moderate in flavour. On the contrary, in small streams running through a fertile soil, fish are frequently killed of a most satisfactory size and weight. In rapid rivers the beds of which are formed of limestone rock, Trout are upon an average, not of a size acceptable to an angler who scouts the idea of a ¼ lb fish. In such rivers they get knocked about very much during heavy floods, and the rapidity with which the streams carry away the feed, either at top or bottom, is against them.

In North Yorkshire and Durham, where many Trouting streams are recipients of the washing of the refuse ore of the lead mines, commonly called hush, fish are not either so plentiful, or near the average size they used to be, when the hush was not so prevalent as it is at present. The hush must certainly be injurious to all kinds of fish, and I think it very probable that the young fry suffer very much from it, even to the extent of being in some instances completely destroyed by it. But there are other causes, independent of hush, &c., why fish are generally smaller in size and number than they used to be in "the days of old." An increasing population has visibly increased the number of anglers, and also of parties making use of most destructive wholesale methods of taking fish, to which any amount of angling is indeed comparatively harmless. Angling clubs conducted with energy and liberality have in some places repressed nefarious practices, and some rivers are coming round again, that previous to the protective system were nearly cleaned out.

The artificial production of Trout and Salmon, has of late years been tried with success. Those who are curious and interested in pisciculture may obtain a pamphlet on the artificial production of fish by Piscarius, published by Reeve & Co., Henrietta Street, Covent Garden, London.

ANGLING IMPEDIMENTS.

The weather may be propitious, the humour of the fish charming, two capital items, that can only now and then be inserted in an angler's diary; but some things may occur to spoil a day's diversion, commenced even under the most favourable auspices; for instance, let us suppose that a man (who whilst "realizing the charms of solitude") is nevertheless carefully and cautiously fishing with success in a clear low water; how great then must be his vexation, and disappointment, when he sees looming in the distance a rod, and net, the owner of which is soon distinctly visible. It does not require a moment's consideration as to what he must now do; he must either give up fishing for that day, or seek some fresh ground, because any person coming fishing down a low water, or even walking close to the banks of it, scares the fish to such an extent, that making for their holds, they will probably remain there for some hours. My object in reference to the above suppositious statement (which many anglers will find too often a reality) is to demonstrate to the inexperienced, what very meagre sport any one must have in a clear, low water, previously fished on the same day.

Reversing the case, that is to say, a day or two after a flood, and when of course there is plenty of water, and also, when fish are not so soon alarmed and disturbed; I hold even then, first come first served, to be the order of the day; for when fish are inclined to feed, any person in advance of you has a decided advantage, and particularly so, should he be either trolling, or worm fishing. In wide rivers however, you may (owing perhaps to a feed coming on) have excellent diversion where a person who has preceded you half an hour, or so, has had but indifferent success. If there is only plenty of water, companionship is admissable, though I am inclined to suppose that (under all circumstances) a solitaire has a decided advantage; for this reason, that two or more persons, get over the ground far too quickly, and do not fish in that true, steady, and careful way, they perhaps would do if alone; just whipping the stream here and there, hurrying over the ground, and so spending probably half their time in walking, instead of fishing; but in free waters, where anglers are sometimes as thick as blackberries, and a man cannot do as he likes, the "go ahead" system often proves the best. Some way or other there is generally some sport to be had in streams, free from hush, but many rivers are daily subject to it, causing great interruption, to say nothing of total stoppage to angling pursuits for many successive days. Slight hushes, when the water is low, are so far serviceable, that by partially discolouring the water, fish take the artificial fly, especially the Black Midge, more boldly than they would do if the water remained clear. Taken altogether, the hush undoubtedly levies a considerable tax on the patience of those anglers who fish in its vicinity.

BARNARD CASTLE AS AN ANGLING STATION.

I beg to offer a few observations to strangers in reference to Barnard Castle as an angling station. The facilities offered by a railway, the beautiful local scenery, the fishing, and the excellent accommodations to be had at reasonable charges, are all attractive considerations for Tourists and Anglers, who will find Barnard Castle a central, pleasant, and convenient place of abode, during any length of time they may please to devote to angling or other recreations. Barnard Castle is particularly well adapted for an angling station; the river Tees is in close proximity to the town, the river Greta distant only about three miles, and there are several other good streams within easy distances.

Gentlemen who obtain leave from W. S. Morritt, Esq., to fish in that portion of the Greta which is strictly preserved, abounding in Trout, and encompassed by those woods and banks alluded to in *Scott's Rokeby*, will find the Inn kept by Mr. Ward, Greta Bridge, very comfortable and convenient. A good day's sport may be had above Bowes; when there happens to be too much water for angling purposes, some few miles lower down.

WEATHER SIGNS AND CHANGES.

Mists. — A white Mist in the evening over a meadow with a river, will be drawn up by the sun next morning, and the day will be bright; five or six Fogs successively drawn up portend rain; when there are lofty hills, and the mist which hangs over the lower lands draws towards the hills in the morning, and rolls up to the top, then it will be fair, but if the mist hangs upon the hills, and drags along the woods, there will be rain.

Clouds. — Against much rain the clouds grow bigger and increase very fast, especially before thunder. When the clouds are formed like fleeces, but dense in the middle and bright towards the edges, with a bright sky, they are signs of frost, with hail, snow or rain. If clouds breed high in the air, with white trains like locks of wool, they portend wind, probably rain. When a generally cloudiness covers the sky, and small black fragments of clouds fly underneath, they are sure signs of rain, and probably it may last some time. Two currents of air always portend rain, and in Summer, thunder.

Dew. — If the dew lies plentifully upon the grass after a fair day, it is a sign of another; if not, and there is no wind, rain must follow. A red evening shews fine weather, but if it spread too far upwards from the horizon in the evening, and especially in the morning, it fortells wind or rain, or both. When the sky in rainy weather is tinged with sea green, the rain will increase; if with blue, it will be showery.

Heavenly Bodies. — A haziness in the air which fades the sun light and makes the orb appear whiteish or ill defined, or at night if the moon and stars grow dim and a ring encircles the former, rain will follow. If the Sun's rays appear like Moses' horn, white at setting or shorn of his rays, or goes down into a bank of clouds in the horizon, bad weather may be expected. If the moon looks pale and dim, rain may be expected; if red, wind; and if her natural colour, with a fair clear sky, fine weather; if the moon is rainy throughout, it will clear at the change, and perhaps the rain return a few days after. If fair throughout, and rain at the change, the fair weather will probably return at the fourth or fifth day.

Wind.—If the wind veers much about, rain is certain; in changing, if it follows the course of the sun, it brings fair weather; the contrary, foul; whistling of the wind is a sure sign of rain.

Meteors.—The Aurora Borealis after warm days is generally succeeded by cooler air; shooting stars are supposed to indicate rain.

Animals.—Before rain, swallows fly low; dogs grow sleepy and eat grass; waterfowl dive much; fish will not bite; flies are more than ordinary troublesome; toads crawl about; moles, ants, bees and insects are very busy; birds fly low for insects; swine, sheep and cattle are uneasy; and it is not without its effect on the human frame.

Weather Table.—The following table, ascribed to Dr. Herschel, and revised by Dr. Adam Clark, constructed upon philosophical consideration of the sun and moon, in their several positions respecting the earth, and confirmed by experience of many years actual observation, furnishes the observer without further trouble, with the knowledge of what kind of weather may be expected to succeed, and that so near the truth, that in a very few instances will it be found to fail.

Observation by Dr. Kirwan.—When there has been no particular storm about the time of the Spring Equinox (March 21st); if a storm arises from the east on or before that day, or if a storm from any point of the compass arise near a week after the Equinox, then in either of these cases the succeeding Summer is generally dry four times in five, but if the storm arises from the S.W. or W.S.W. on or just before the Spring Equinox, then the Summer following is generally wet five times in six.

WEATHER TABLE.

NEW & FULL MOON.	IN SUMMER.	IN WINTER.
If it be New or Full Moon, or the Moon entering into the first or last quarter at 12 at noon or between 12 and 2	Very Rainy	Snow and rain
2 and 4 in the Afternoon	Changeable	Fair and Mild
4 and 6 Evening	Fair	Fair
6 and 8	Fair if wind at N West, Rainy if S, or S. West	Fair and Frosty, if wind at North or N. East, Rain or Snow, if South or S. West
8 and 10	Ditto	Ditto
10 and 12 Night	Fair	Fair and Frosty
12 and 2 Morning	Ditto	Hard frost unless wind South or S. West
2 and 4 Morning	Cold with frequent showers	Snow and Storm
4 and 6 Morning	Rain	Ditto
6 and 8 Morning	Wind and Rain	Stormy Weather
8 and 10 Morning	Changeable	Cold Rain, if wind be West, Snow if East
10 and 12 Morning	Frequent Showers	Cold with high wind

NOTICES OF RARE AND CURIOUS ANGLING BOOKS.

There exists a very rare and remarkable work, "*A Book of Angling or Fishing, wherein is shewed by conference with Scriptures, the agreement between the Fisherman, Fishes, and Fishing of both natures, spirituall and temporall, by Samuel Gardner, Doctor of Divinitie.*" — "I will make you fishers of men." — Matt. iv. 19. London, printed for Thomas Pinfoot, 1606.

Walton tells the honest angler that the writing of his book was the recreation of a recreation; his motto on the title page of his book was, "Simon Peter said let us go a fishing, and they said we also will go with thee" — John xxi. 3. This passage is not in all the editions of the *Complete Angler*, but was engraven on the title page of the first edition, printed in 1653.

Advertisement of Walton's angler, 1653. There is published a book of eighteenpence price, called "*The Compleat Angler, or Contemplative Man's Recreation, being a Discourse of Fish and Fishing, not unworthy the perusal.*"

These works may now be considered as great bibliomaniacal curiosities.

ADDENDA.

It is altogether a mistake to suppose that large flies are required for large rivers; on the contrary, with the exception of the Palmers, small hackle flies will be found to answer best, these, together with the Black, Blue and Dun Midges, (Spring and Autumn excepted), have a decided advantage in general over dubbed or hackle winged flies. In small brooks after a flood, winged flies often kill well, those with Orange, Black, Crimson, and Yellow bodies are the best. Grass Hoppers, the Cabbage Caterpillar, the Breccan or Fern Clock, will all take Trout; but as there are other natural baits to be had at the time these are in season, which I have noted, and which are more to be depended upon, I have not given any special instructions as to the use of the above. The Grass Hopper and Caterpillar are tiresome baits to fish with, and more a matter of fancy than utility; the Breccan Clock found amongst fern, fished like the May-fly is the best of the lot, and at times kills pretty well. Having made no allusion in my work to Lake or Pond Fishing, I may now observe, that four flies upon a stretcher, one yard apart from each other, are sufficient for Ponds. On Lakes, fishing from a boat, you may have six or eight, or even more flies upon a stretcher. In Lake and Pond fishing, the Palmers and large winged flies are the best, particularly when there is a good curl upon the water; but when there is no wind stirring, the small hackle or very small winged flies will, as regards Ponds, be frequently found to kill much better than larger flies, particularly in mornings and evenings during Summer. As fly fishing and trolling are the only reliable angling means and devices for taking Trout in Lakes and Ponds, I have nothing further to add, than that a good rod and sound tackle are essential requisites.

FINIS.

FOOTNOTES

[1] This fish has only been observed in the Tees during the last few years.

[2] Very many clever men have written diffusely on Ichthyology. Aristotle was one of the first who divided fishes into different orders, he divided them into three, but Linnæus separated them into five.

[3] Under favourable circumstances you may begin to troll with the Minnow about the middle of the month.

[4] The Duns are first-class flies all the season, beginning with the Blue Dun in March and April,--The Yellow Dun, little Iron Blue Dun on cold windy days,--July Dun Cut, Blue Gnat, and Willow Fly.

[5] A good Fly on cold windy days.

[6] This fly kills well when the water is low and fine.

[7] When there is much water some of the Spring and Autumn Hackle flies may be dressed on No. 3 Hooks.

[8] The water beetle is chiefly instrumental in conveying the spawn of various kinds of fish to waters, where such species had previously been unknown.